Cosmin Radu Popa

Current-Mode Analog Nonlinear Function Synthesizer Structures

Springer

Cosmin Radu Popa
Faculty of Electronics, Telecommunications
and Information Technology
University Politehnica of Bucharest
Bucharest
Romania

ISBN 978-3-319-01034-2 ISBN 978-3-319-01035-9 (eBook)
DOI 10.1007/978-3-319-01035-9
Springer Cham Heidelberg New York Dordrecht London

Library of Congress Control Number: 2013941505

Printed on acid-free paper

Springer is part of Springer Science+Business Media (www.springer.com)

This book is dedicated to
my Grandmother

Preface

Analog signal processing structures represent an important part of analog VLSI designs, a multitude of specific techniques being developed in order to maximize their performances. Comparing with digital structures that respond to the same requirements, the analog approach of signal processing presents the advantages of a real-time operation and of the possibility of their facile integration, especially for portable devices. The development of specific superior-order approximation functions that constitutes the functional cores of the function synthesizer circuits presented in this book increases the flexibility of designs, making them comparable, from this point of view, with latest digital computational structures. The design of analog signal processing computational structures from the perspective of multifunctionality (the utilization of a single functional core for generating a large number of continuous functions) additionally reduces both power consumption and silicon area per implemented function.

The aim of this book is to propose a large diversity of analog function synthesizer structures, based on accurate approximation functions, that are able to implement a multitude of circuit functions, with many applications in analog signal processing: exponential, Gaussian, or hyperbolic functions. Generalizing the methods for obtaining these particular functions, the book analyzes superior-order approximation functions, which represent the core for developing CMOS analog nonlinear function synthesizers.

The method presented for generating continuous functions is based on the utilization of superior-order approximation functions. Comparing with the classical method of designing analog function synthesizer circuits using the limited Taylor series expansion of the generated function, the utilization of high-accuracy approximation functions permits an important reduction of the overall complexity of designed computational circuits. In order to maximize the performance/complexity ratio, some fundamental requirements must be taken into account when developing the general form of an approximation function. First, technical considerations referring to the CMOS implementation of these approximation functions impose a particular structure of primitive mathematical functions that compose them. Because the simplest and most performance basic CMOS computational building blocks are represented by the squaring and multiplying/dividing circuits, the fractional and the squaring functions represent the most convenient primitive mathematical functions for developing any superior-order approximation function. Second, the requirements for an increased accuracy of approximations impose an independence of the output variable on technological parameters, from this point of view the current-mode operation of the basic building blocks representing the best possible choice. Additionally, the current-mode operation contributes to a reduction of the minimal supply voltage and to an important improving of the computational structures' frequency response. In most cases, the general form of the approximation function is represented by a linear combination of a finite number of primitive functions (usually smaller or equal with two) and some additional terms (linear or polynomial terms). The particular forms of these superior-order approximation functions are developed considering the specific type of the approximated function (odd or even functions).

A general tradeoff that must be taken into account is referring to relation between the computational structures' accuracy and complexity and their capability of generating a multitude of continuous functions. The increasing of the number of computed functions increases also the complexity of the function synthesizers and reduces the possible accuracy that can be obtained using a fixed approximation function. On the other hand, the grouping of approximated functions in some specific classes and the development of analog function synthesizer circuits dedicated only to one specific class maximize the performances of computational structures and decrease the complexity of their implementation.

The first chapter presents possible realizations of exponential function synthesizer circuits, developing third-order and fourth-order approximation functions in order to design accurate computational structures. The circuits are structured considering the order of the approximation functions they are based on and illustrating the block diagrams and the concrete implementations of these structures. Exponential function synthesizer circuits are evaluated from the point of view of their accuracy in analytical and graphical manner, for each particular computational structure being evaluated its output dynamic range. A new general method for increasing the exponential function synthesizers' output dynamic range is presented, the reducing of the effective variation range of the input variable using proper variable changing contributing to fulfill this desiderate.

High-accuracy superior-order approximation functions are developed in the second chapter for generating the Gaussian function. The particular form of the approximation functions presented in this chapter represents a consequence of some important particularities of the Gaussian function (which is even function that allows facile variable changing in order to increase the output dynamic range of computational circuits that generate it). The chapter analyzes a multitude of possible realizations of Gaussian function synthesizer circuits, based on superior-order particular approximation functions: fourth-order, sixth-order, and eighth-order approximation functions. In order to improve the area of operation of developed Gaussian circuits, convenient variable changing are considered for each analyzed approximation function. Analytical and graphical analyses are performed for determining the performances of these superior-order approximation functions.

The analysis, design, and performances' optimization of hyperbolic functions represent the goal of the third chapter. The nonlinear function synthesizer circuits, developed for generating the hyperbolic sinusoidal, co-sinusoidal, and tangent functions use specific superior-order approximation functions. The form of these approximation functions is strongly influenced by the particularities of hyperbolic functions, being correlated with the requirements of minimizing the computational circuits' complexity.

The fourth chapter analyzes the possibility to realize general analog function synthesizer structures, designed for implementing a multitude of circuit functions. Comparing with classical designs in which for any particular function it is designed a specific circuit, the re-utilization of the same functional core for generating a large number of circuit functions presents important advantages. As the most important part of the power is consumed by the functional core and because the largest complexity and design efforts are concentrated for designing this part of the circuit, the utilization of a single computational structure for generating a multitude of circuit functions removes all redundant parts of the circuits and, in consequence, strongly reduces the power consumption and the required complexity per implemented function. The second-order approximation is not able to approximate with enough accuracy a continuous function, resulting the necessity of increasing the order of approximation for improving the analog function synthesizer' performances. In the context of the equilibrium that must be considered between the accuracy of computational circuits and the complexity of their CMOS implementation, the third-order of the approximation represents, for most of continuous functions, a convenient choice. Similar technological restrictions like in previous chapters must be considered for the implementation in CMOS technology of these third-order approximation functions.

For applications that require a more accurate approximation of the circuit functions, the fourth-order approximation is usually enough accurate for generating the most important circuit functions, from the perspective of their applications in analog signal processing and of fulfilling the conditions imposed by these applications. Comparing with previously presented computational structures based on third-order approximation functions, the same requirements must be taken into account when designing a high-precision fourth-order analog function synthesizer circuit.

Contents

Chapter 1
Wide Output Dynamic Range Exponential Function Synthesizers

1.1 Introduction

Exponential functions present a multitude of applications in VLSI designs, such as telecommunication applications, medical equipments, disk drives, hearing aids, neural networks, neural algorithms, neuro-fuzzy and classification applications, pattern recognition, on-chip unsupervised learning, or wavelet transforms. The requirements related to the frequency response of computational structures that implement the exponential function impose important limitations to their CMOS realizations. In this sense, the exponential characteristic of the subthreshold-operated MOS active devices cannot be used, because the extremely low values of drain currents in this region strongly limit the frequency response of the designed circuits. The single choice is represented by the biasing in saturation of MOS transistors that compose the exponential function synthesizers, exploiting the approximately squaring characteristic of the active devices in this region.

A very important requirement of an exponential function synthesizer is related to the output dynamic range that could be obtained for a high-accuracy generating of the exponential function. Additionally, the complexity of CMOS circuit that implements the function synthesizer circuit must be decreased using specific design techniques. In this context, some particularities of developing the computational architectures can be considered. First, the current-mode operation of the function synthesizer usually removes (in a first-order analysis) the dependence of the circuit operation on technology and on temperature-caused errors. This error reduction allows a decreasing of the order of approximation for the designed exponential function synthesizer circuit, contributing to an important decreasing of the overall complexity of the computational structure. Second, the blocks that implement the approximation functions must be chosen from the circuits having the lowest complexity of their implementations in CMOS technology. Exploiting the squaring characteristic of MOS transistors biased in saturation region, the simplest computational structures are represented by the current-mode squaring and multiplier/divider circuits (MDs).

C. R. Popa, *Current-Mode Analog Nonlinear Function Synthesizer Structures*,
DOI: 10.1007/978-3-319-01035-9_1,

Taking into account the previous general considerations, accurate exponential function synthesizer circuits could be developed, considering a trade-off between the accuracy and the silicon area of the computational structure, for a given CMOS technology. Using specific design techniques, the developed computational structures will present very wide output dynamic ranges comparing with classical designs.

There will be analyzed in this chapter a multitude of possible realizations of exponential function synthesizer circuits, based on third-order and fourth-order approximation functions. The presented computational circuits will be structures following the order of the approximation function they are based on, illustrating the block diagrams and the concrete implementations of these structures. The accuracies of exponential function synthesizer circuits are evaluated both in analytical and graphical manner, determining, for each particular computational structure, the output dynamic range.

1.2 General Considerations About Exponential Function Synthesizers

Third-order and fourth-order general approximation functions that are presented in this chapter will be particularized for implementing the important class of exponential functions. Additionally, in order to improve the output dynamic ranges of developed exponential function synthesizer circuits, specific variable changing will be presented. The increasing of the output dynamic ranges of the computational structures using this method exploits the possibility of reduction in the equivalent variation range for the input variable. This is equivalent to the biasing of exponential function synthesizer circuits in the neighborhood of their static operating point, this fact having as consequence the increasing of their overall accuracy and, thus, of their output dynamic range.

1.2.1 General Variable Changing

The particular variable changing that can be used for increasing the output dynamic range of exponential function synthesizer circuits is correlated with the practical considerations imposed by their CMOS implementation. Taking into account these restrictions, the general variable changing can be expressed as follows:

$$x \rightarrow \frac{x}{\alpha}, \tag{1.1}$$

α parameter being imposed by a trade-off between the output dynamic range of the exponential computational circuit and the complexity of its CMOS implementation. The following mathematical identity will represent the functional basis for increasing the output dynamic range of the function synthesizer:

$$\exp(x) = \left[\exp\left(\frac{x}{\alpha}\right)\right]^{\alpha}. \tag{1.2}$$

The value of α parameter is also given by a trade-off between the circuit complexity and its output dynamic range. Additionally, the technological context imposes other restrictions for the value of this parameter. Most concrete, its value will be correlated with the possibility of a facile implementation in CMOS technology of x^{α} function. Because the most convenient realization of a CMOS computational structure using MOS active devices biased in saturation region is represented by the squaring function and, with a reasonable additionally complexity increasing, by the x^4 functions, the practical usual variable changing can be considered to be $\alpha = 2$ and $\alpha = 4$. So, the variable changing will be $x \to x/2$ and $x \to x/4$, the functional relation (1.2) becoming:

$$\exp(x) = \left[\exp\left(\frac{x}{2}\right)\right]^{2}, \tag{1.3}$$

or

$$\exp(x) = \left[\exp\left(\frac{x}{4}\right)\right]^{4}. \tag{1.4}$$

On theoretical level, it is possible to choose others variable changing (for example, $x \to x/6$), but they will require a relatively large complexity of function synthesizers they will be based on. In conclusion, it is preferable to maintain the previous variable changing and to increase the accuracy of the approximation functions that implement the $\exp(x/2)$ or $\exp(x/4)$ functions by increasing their form or order of approximation.

1.2.2 Choosing the Order of Approximation for the Fundamental Approximation Function

Design of high-accuracy exponential function synthesizer circuits with wide output dynamic range is based on two important things. The first one is represented by the previously analyzed variable changing. The second one is based on an accurate approximation of the fundamental function that generates particular forms of the exponential function ($\exp(x/2)$ or $\exp(x/4)$ functions, respectively). In order to implement these mathematical functions, high-accuracy approximation functions must be developed. The problems that appear are related to the order of approximation that is convenient to be considered for these approximation functions in order to fulfill all the accuracy requirements, but with consuming minimal hardware resources for implementing the exponential function synthesizers.

The most important goals of designing exponential function synthesizers are related to theirs accuracies and with their output dynamic ranges. The output dynamic range is correlated with the maximal range of the x input variable and with the maximal accuracy that is accepted for the designed synthesizer circuit. For example, a maximal range for x variable between 0 and 6 is equivalent to a maximal output dynamic range of the exponential function synthesizer of $20\lg[\exp(8)]\text{dB} \cong 52$ dB. An increasing of the x range to 0–8 domain increases the output dynamic range to approximately 70 dB, while an additional increasing of the x variable range to 0–10 domain allows to obtain a maximal dynamic range of the synthesizer circuit of approximately 87 dB. In conclusion, in correlation with the maximal output dynamic range that is imposed to the exponential function synthesizer circuit and, also, with the minimal accuracy that is accepted for this circuit, the minimal order for the approximation function can be determined. For the exponential function approximation, second-order, third-order, and fourth-order approximations for the fundamental exponential function are usually enough for most of the practical applications. These reasonable values of the order of approximations are consequences of the variable changing the developed exponential function synthesizers are based on.

A continuous $f(x)$ function can be expanded in Taylor series using the following polynomial function:

$$f(x) = m + \text{nx} + \text{px}^2 + \text{qx}^3 + \text{rx}^4 + \text{sx}^5 + \text{tx}^6 + \ldots, \tag{1.5}$$

m, n, p, q, r, s, and t being constant parameters of the expansion, imposed by the particular function $f(x)$. The exponential function can be expand in Taylor series as follows:

$$\exp(x) = 1 + x + \frac{x^2}{2} + \frac{x^3}{6} + \frac{x^4}{24} + \frac{x^5}{120} + \frac{x^6}{720} + \ldots. \tag{1.6}$$

An nth order approximation of the exponential function is equivalent to the identity of the first $n + 1$ terms of Taylor series expansions of exponential function and approximation function.

1.2.3 *CMOS Implementations of Exponential Function Synthesizers' Fundamental Blocks*

Considering the requirements of facile implementation in CMOS technology and of improved accuracy, current-mode squaring and MDs represent the most convenient realizations of fundamental blocks for developing performance exponential function synthesizers, with minimal hardware resources consumption.

1.2.3.1 CMOS Implementation of the Current-Mode Squaring Circuit

In order to reduce the complexity of the exponential function synthesizers and to increase their overall accuracy, the current-mode squaring circuit (SQ) presented in Fig. 1.1 will be used:

The functional relation that characterizes the operation of the SQ from Fig. 1.1 is

$$I_G = \frac{I_F}{4} - \frac{I_E}{2} + \frac{I_E^2}{4I_F}. \tag{1.7}$$

The symbolical representation of the SQ from Fig. 1.1 is presented in Fig. 1.2.

1.2.3.2 CMOS Implementation of the Current-Mode Multiplier/Divider Circuit

A possible realization of the current-mode MD, required by the implementation of exponential synthesizers, is presented in Fig. 1.3 [1], while its symbolical representation is shown in Fig. 1.4.

The I and I' currents from Fig. 1.3 can be expressed as follows:

$$I = \frac{I_C}{4} - \frac{I_A - I_B}{2} + \frac{(I_A - I_B)^2}{4I_C} \tag{1.8}$$

and

$$I' = \frac{I_C}{4} - \frac{I_A + I_B}{2} + \frac{(I_A + I_B)^2}{4I_C} \tag{1.9}$$

resulting

$$I_D = I' - I = -I_B + \frac{I_A I_B}{I_C} \tag{1.10}$$

Fig. 1.1 SQ implementation

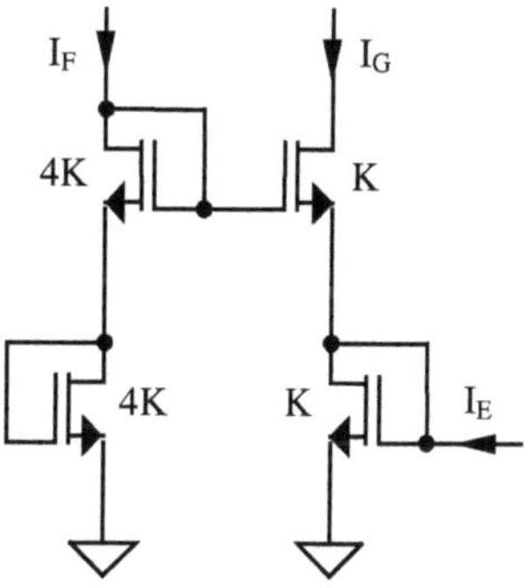

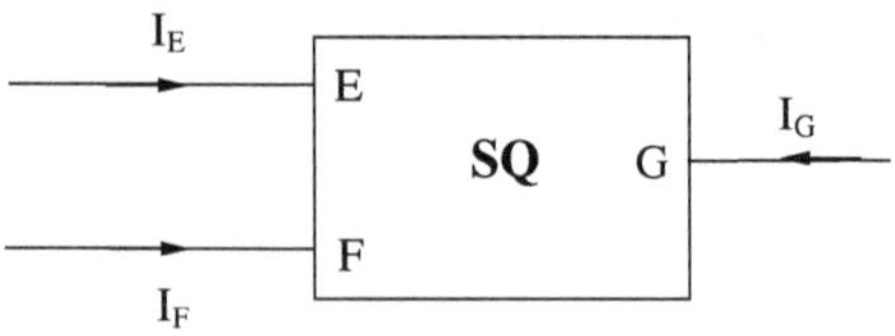

Fig. 1.2 Symbolical representation of the SQ

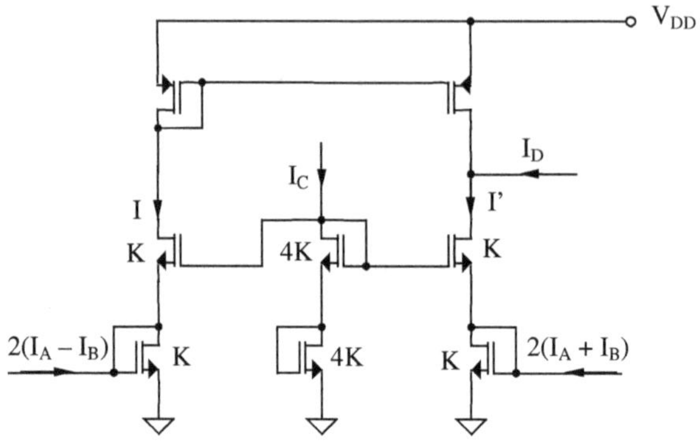

Fig. 1.3 MD implementation [1]

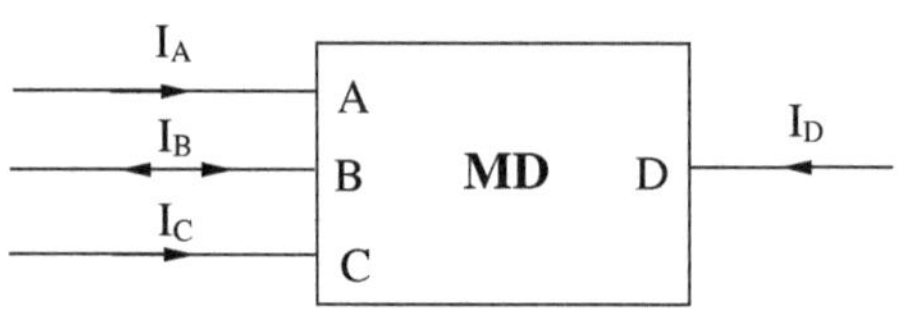

Fig. 1.4 Symbolical representation of the MD

1.3 Wide Output Dynamic Range Exponential Function Synthesizers Based on Second-Order Approximation Functions

1.3.1 First Class of Wide Output Dynamic Range Exponential Function Synthesizers with Second-Order Approximation

1.3.1.1 Approximation Function Without Variable Changing

Considering the previous design restrictions referring to the minimization of the exponential function synthesizer complexity and to the increasing of its overall accuracy, the first general form of the second-order approximation function can be expressed as follows:

$$g_{1a}(x) = \frac{b}{1 + \text{ax}} + c \quad (1.11)$$

a, *b*, and *c* being constant coefficients. Their particular values can be determined imposing the condition that the second-order Taylor series of $g_{1a}(x)$ function to be identical with the third-order series of $f(x) = \exp(x)$ approximated function. There are three coefficients of the general expression of the approximation function because it is necessary to have an identity between the first three terms of $f(x) = \exp(x)$ and $g_{1a}(x)$ functions (required by the second-order approximation). This condition imposes the following particular form of $g_{1a}(x)$ function:

$$g_{1a}(x) = \frac{n^2}{p} \frac{1}{1 - \frac{p}{n}x} + \left(m - \frac{n^2}{p}\right) \quad (1.12)$$

Replacing the *m*, *n*, and *p* coefficients from the Taylor series expansion associated with the exponential function, it results

$$g_{1a}(x) = \frac{4}{2 - x} - 1. \quad (1.13)$$

For small values of *x* variable, the approximation error of $f(x) = \exp(x)$ function using $g_{1a}(x)$ function can be approximated with the third-order error:

$$\varepsilon_{f(x)}^{g_{1a}(x)} \cong \frac{x^3}{12\exp(x)}. \quad (1.14)$$

If the *x* input variable is increasing, more terms having an order greater or equal with three must be considered. A comparison between $g_{1a}(x)$ approximation function and $f(x) = \exp(x)$ function is presented in Table 1.1. The graphical representations of $g_{1a}(x)$ and $f(x) = \exp(x)$ functions are shown in Fig. 1.5, while the graphical representation of the approximation error, $\varepsilon(x)$, defined as the difference between $g_{1a}(x)$ and $f(x) = \exp(x)$ functions, is presented in Fig. 1.6.

The block diagram of the exponential function synthesizer circuit based on $g_{1a}(x)$ approximation function is presented in Fig. 1.7.

Table 1.1 Comparison between $g_{1a}(x)$ approximation function and $f(x) = \exp(x)$ function

x	$f(x)$ (dB)	$g_{1a}(x)$ (dB)	ε (dB)
−1.4	−12.17	−15.07	−2.90
−1.2	−10.43	−12.04	−1.61
−0.8	−6.95	−7.36	−0.41
−0.4	−3.48	−3.52	−0.04
0.0	0.00	0.00	0.00
0.4	3.48	3.52	0.04
0.8	6.95	7.36	0.41
1.2	10.43	12.04	1.61
1.4	12.17	15.07	2.90

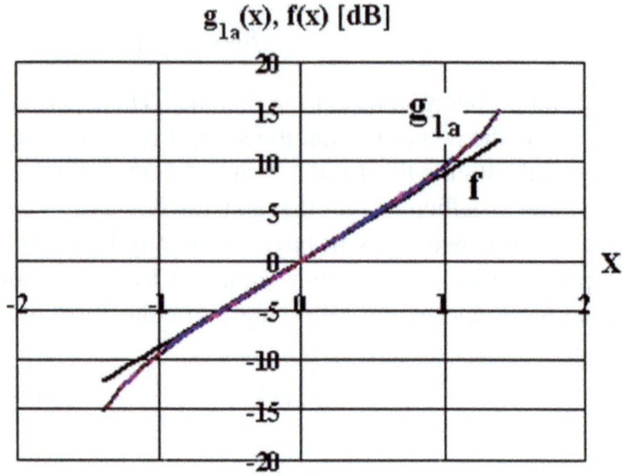

Fig. 1.5 Graphical representation of $g_{1a}(x)$ and $f(x) = \exp(x)$ functions

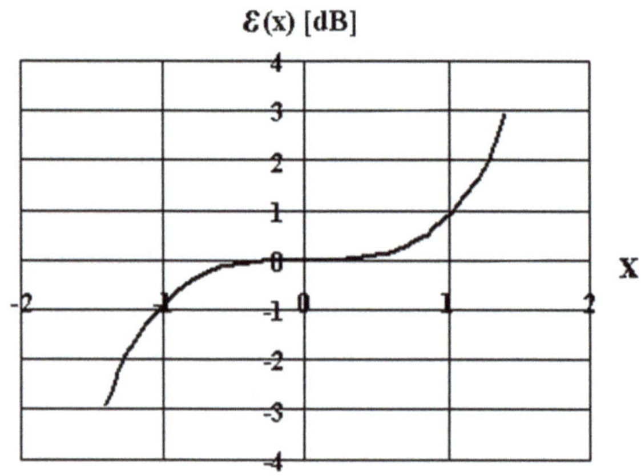

Fig. 1.6 Graphical representation of the approximation error, $\varepsilon(x)$

The expression of I_G current is

$$I_G = \frac{1}{4}(2I_O - I_{\mathrm{IN}}) - 2I_O + \frac{16I_O^2}{4(2I_O - I_{\mathrm{IN}})}, \tag{1.15}$$

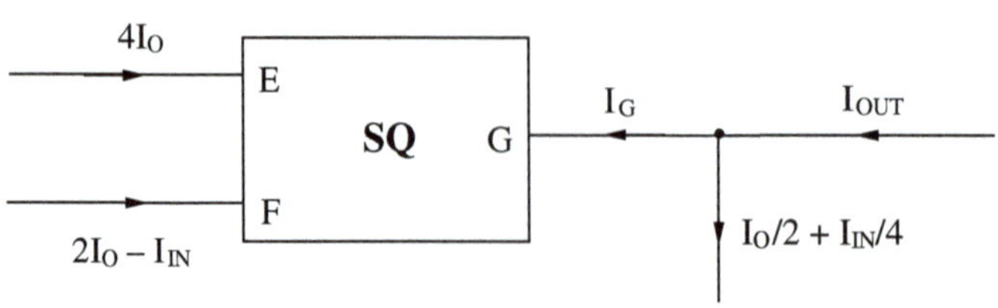

Fig. 1.7 Block diagram of the exponential function synthesizer circuit based on $g_{1a}(x)$ approximation function

resulting the following expression of I_{OUT} current:

$$I_{OUT} = I_G + \frac{1}{4}I_{IN} + \frac{1}{2}I_O = \frac{16I_O^2}{4(2I_O - I_{IN})} - I_O, \tag{1.16}$$

equivalent to

$$I_{OUT} = I_O \left[\frac{4}{2 - \left(\frac{I_{IN}}{I_O}\right)} - 1 \right] \tag{1.17}$$

So, I_{OUT} current approximates the exponential function using $g_{1a}(x)$ approximation function:

$$I_{OUT} = I_O g_{1a}\left(\frac{I_{IN}}{I_O}\right) \cong I_O \exp\left(\frac{I_{IN}}{I_O}\right). \tag{1.18}$$

The CMOS implementation of the exponential function synthesizer circuit based on $g_{1a}(x)$ approximation function is presented in Fig. 1.8 [1].

1.3.1.2 Approximation Function with First Variable Changing

In order to increase the maximal output dynamic range of the exponential function synthesizer based on $g_{1a}(x)$ approximation function, one of the previous presented variable changing can be used. The first variable changing that can be implemented in CMOS technology with a reasonable increasing of the circuit complexity is $x \to x/2$, resulting the following improved approximation function:

$$g_{1b}(x) = \left(\frac{8}{4 - x} - 1\right)^2. \tag{1.19}$$

A comparison between $g_{1b}(x)$ approximation function and $f(x) = \exp(x)$ function is presented in Table 1.2. The graphical representations of $g_{1b}(x)$ and

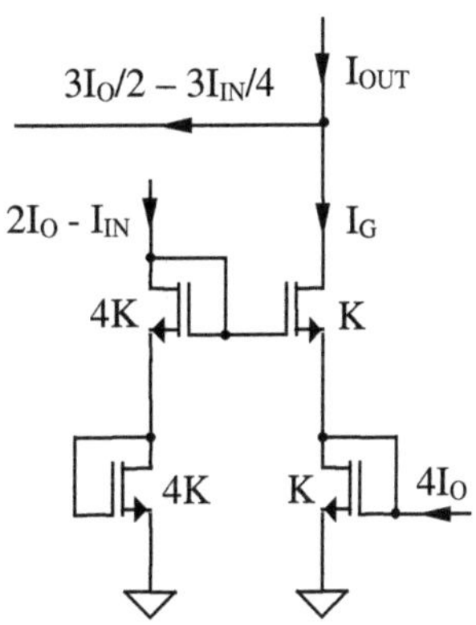

Fig. 1.8 CMOS implementation of the exponential function synthesizer circuit based on $g_{1a}(x)$ approximation function [1]

Table 1.2 Comparison between $g_{1b}(x)$ approximation function and $f(x) = \exp(x)$ function

x	$f(x)$ (dB)	$g_{1b}(x)$ (dB)	ε (dB)
−1.6	−13.91	−14.72	−0.81
−1.2	−10.43	−10.75	−0.32
−0.8	−6.95	−7.04	−0.09
−0.4	−3.48	−3.49	−0.01
0.0	0.00	0.00	0.00
0.4	3.48	3.49	0.01
0.8	6.95	7.04	0.09
1.2	10.43	10.75	0.32
1.6	13.91	14.72	0.81
2.0	17.38	19.08	1.70

$f(x) = \exp(x)$ functions are shown in Fig. 1.9, while the graphical representation of the approximation error, $\varepsilon(x)$, defined as the difference between $g_{1b}(x)$ and $f(x) = \exp(x)$ functions, is presented in Fig. 1.10.

The block diagram of the exponential function synthesizer circuit based on $g_{1b}(x)$ approximation function is presented in Fig. 1.11.

The expression of I_{Ga} current is

$$I_{Ga} = \frac{1}{4}\left(2I_O - \frac{I_{IN}}{2}\right) - 2I_O + \frac{16I_O^2}{4\left(2I_O - \frac{I_{IN}}{2}\right)}. \tag{1.20}$$

So, I_{OUT}' current will have the following expression:

$$I_{OUT}' = I_{Ga} + \frac{I_O}{2} + \frac{I_{IN}}{8} = \frac{8I_O^2}{4I_O - I_{IN}} - I_O, \tag{1.21}$$

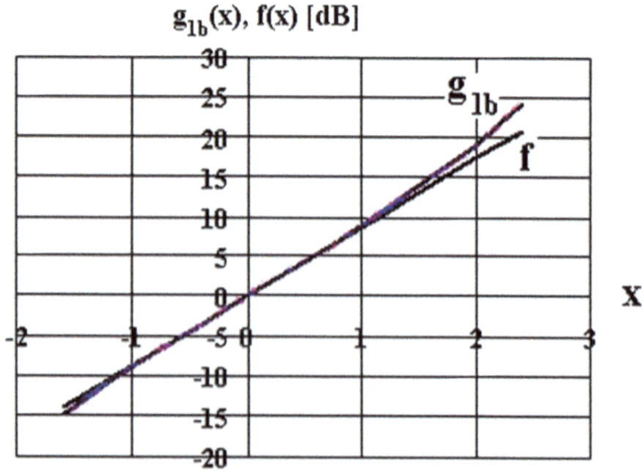

Fig. 1.9 Graphical representation of $g_{1b}(x)$ and $f(x) = \exp(x)$ functions

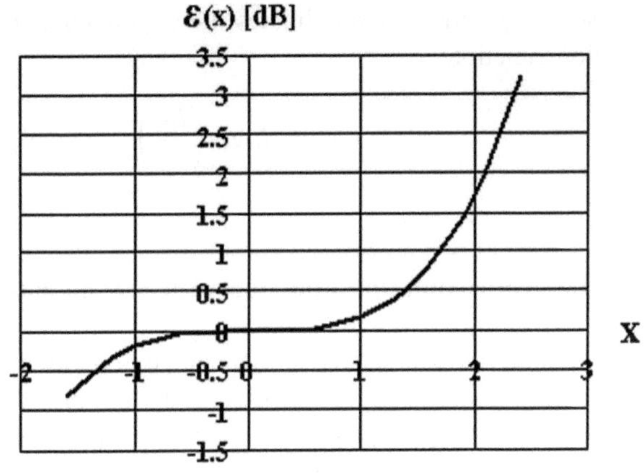

Fig. 1.10 Graphical representation of the approximation error, $\varepsilon(x)$

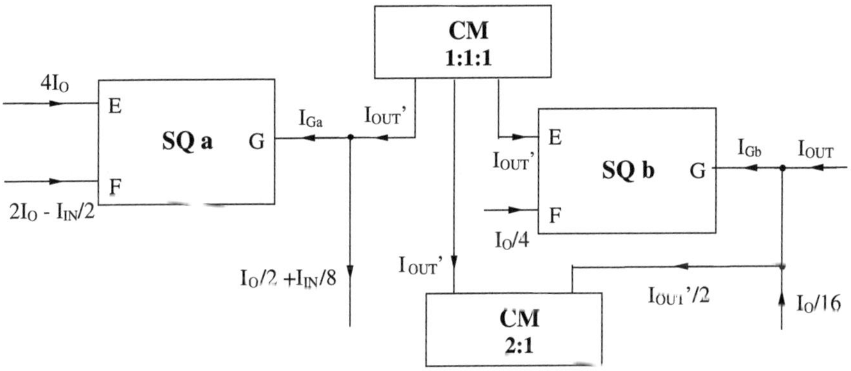

Fig. 1.11 Block diagram of the exponential function synthesizer circuit based on $g_{1b}(x)$ approximation function

equivalent to

$$I_{\mathrm{OUT}}{}' = I_O\left[\frac{8}{4-\left(\frac{I_{\mathrm{IN}}}{I_O}\right)}-1\right] \tag{1.22}$$

The I_{Gb} current can be expressed as follows:

$$I_{\mathrm{Gb}} = \frac{I_O}{16} - \frac{I_{\mathrm{OUT}}{}'}{2} + \frac{(I_{\mathrm{OUT}}{}')^2}{I_O}, \tag{1.23}$$

while the expression of I_{OUT} current is

$$I_{\mathrm{OUT}} = I_{\mathrm{Gb}} - \frac{I_O}{16} + \frac{I_{\mathrm{OUT}}{}'}{2} = I_O\left[\frac{8}{4-\left(\frac{I_{\mathrm{IN}}}{I_O}\right)}-1\right]^2. \tag{1.24}$$

So, I_{OUT} current approximates the exponential function using $g_{1b}(x)$ approximation function:

$$I_{OUT} = I_O g_{1a}\left(\frac{I_{IN}}{I_O}\right) \cong I_O \exp\left(\frac{I_{IN}}{I_O}\right). \tag{1.25}$$

1.3.1.3 Approximation Function with Second Variable Changing

The second variable changing that can be implemented in CMOS technology with a reasonable additional increasing of the circuit complexity, but which strongly increases the output dynamic range of the exponential function synthesizer circuit is $x \to x/4$. It results the following improved approximation function:

$$g_{1c}(x) = \left(\frac{16}{8-x} - 1\right)^4. \tag{1.26}$$

The comparison between $g_{1c}(x)$ approximation function and $f(x) = \exp(x)$ function is shown in Table 1.3, while the graphical representations of $g_{1c}(x)$ and $f(x) = \exp(x)$ functions are presented in Fig. 1.12. The graphical representation of the approximation error, $\varepsilon(x)$, defined as the difference between $g_{1c}(x)$ and $f(x) = \exp(x)$ functions, is shown in Fig. 1.13.

The block diagram of the exponential function synthesizer circuit based on $g_{1c}(x)$ approximation function is presented in Fig. 1.14.

Table 1.3 Comparison between $g_{1c}(x)$ approximation function and $f(x) = \exp(x)$ function

x	$f(x)$ (dB)	$g_{1c}(x)$ (dB)	ε (dB)
−3.6	−31.29	−33.68	−2.39
−3.2	−27.81	−29.44	−1.63
−2.8	−24.34	−25.39	−1.05
−2.4	−20.86	−21.51	−0.65
−2.0	−17.38	−17.75	−0.37
−1.6	−13.91	−14.09	−0.18
−1.2	−10.43	−10.50	−0.07
−0.8	−6.95	−6.97	−0.02
−0.4	−3.48	−3.48	0.00
0.0	0.00	0.00	0.00
0.4	3.48	3.48	0.00
0.8	6.95	6.97	0.02
1.2	10.43	10.50	0.07
1.6	13.91	14.09	0.18
2.0	17.38	17.75	0.37
2.4	20.86	21.51	0.65
2.8	24.34	25.39	1.05
3.2	27.81	29.44	1.63
3.6	31.29	33.68	2.39

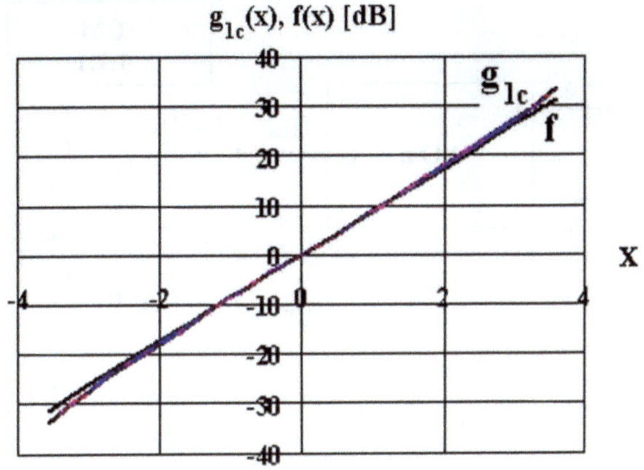

Fig. 1.12 Graphical representation of $g_{1c}(x)$ and $f(x) = \exp(x)$ functions

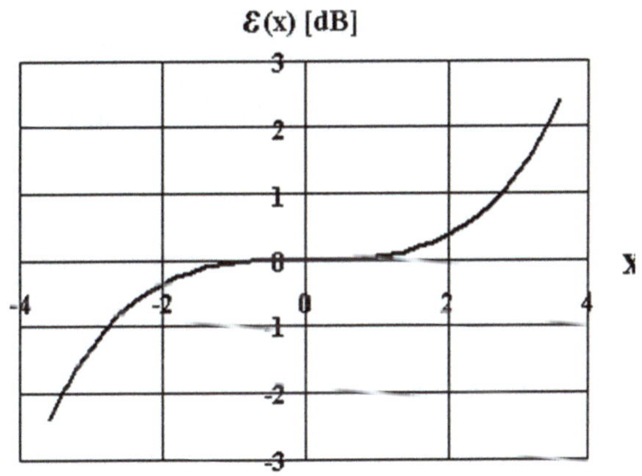

Fig. 1.13 Graphical representation of the approximation error, $\varepsilon(x)$

The expression of I_{Ga} current is

$$I_{\mathrm{Ga}} = \frac{1}{4}(8I_O - I_{\mathrm{IN}}) - 4I_O + \frac{16I_O^2}{8I_O - I_{\mathrm{IN}}}. \quad (1.27)$$

So, $I_{\mathrm{OUT}}{}'$ current will have the following expression:

$$I_{\mathrm{OUT}}{}' = I_{\mathrm{Ga}} + \frac{I_{\mathrm{IN}}}{4} + I_O = \frac{16I_O^2}{8I_O - I_{\mathrm{IN}}} - I_O, \quad (1.28)$$

equivalent to

$$I_{\mathrm{OUT}}{}' = I_O\left[\frac{16}{8 - \left(\frac{I_{\mathrm{IN}}}{I_O}\right)} - 1\right]. \quad (1.29)$$

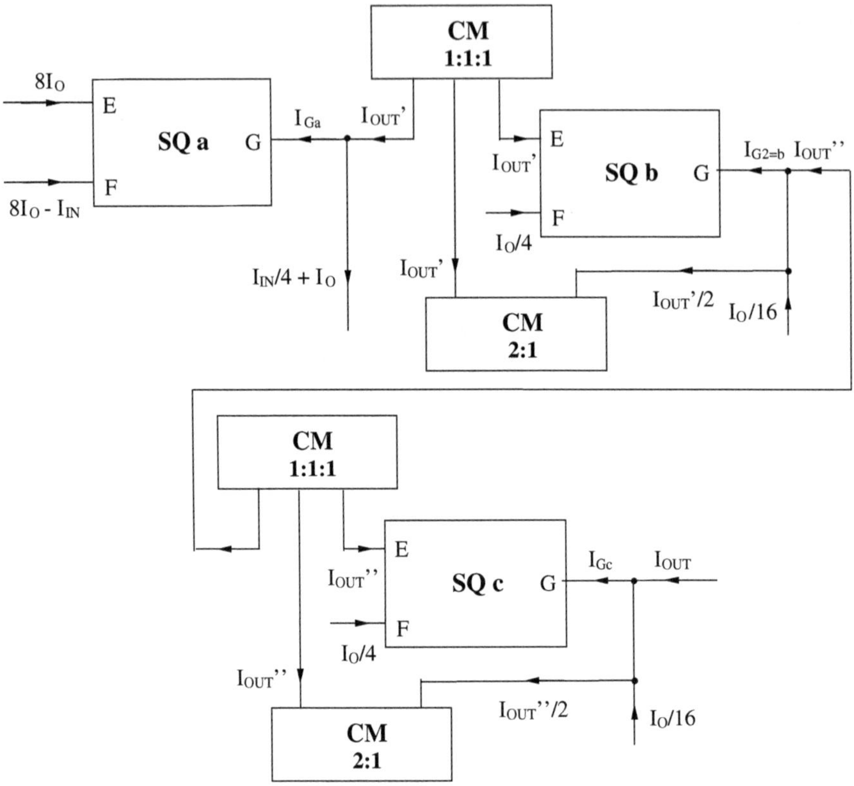

Fig. 1.14 Block diagram of the exponential function synthesizer circuit based on $g_{1c}(x)$ approximation function

The I_{Gb} current can be expressed as follows:

$$I_{\mathrm{Gb}} = \frac{I_O}{16} - \frac{I_{\mathrm{OUT}}'}{2} + \frac{(I_{\mathrm{OUT}}')^2}{I_O}, \tag{1.30}$$

while the expression of I_{OUT}'' current is

$$I_{\mathrm{OUT}}'' = I_{\mathrm{Gb}} - \frac{I_O}{16} + \frac{I_{\mathrm{OUT}}'}{2} = I_O \left[\frac{16}{8 - \left(\frac{I_{\mathrm{IN}}}{I_O} \right)} - 1 \right]^2. \tag{1.31}$$

The expression of I_{Gc} current will be

$$I_{\mathrm{Gc}} = \frac{I_O}{16} - \frac{I_{\mathrm{OUT}}''}{2} + \frac{(I_{\mathrm{OUT}}'')^2}{I_O}, \tag{1.32}$$

resulting the following expression of I_{OUT} current:

$$I_{\mathrm{OUT}} = I_{\mathrm{Gc}} - \frac{I_O}{16} + \frac{I_{\mathrm{OUT}}''}{2} = I_O \left[\frac{16}{8 - \left(\frac{I_{\mathrm{IN}}}{I_O}\right)} - 1 \right]^4. \tag{1.33}$$

So, I_{OUT} current approximates the exponential function using $g_{1b}(x)$ approximation function:

$$I_{\mathrm{OUT}} = I_O g_{1c}\left(\frac{I_{\mathrm{IN}}}{I_O}\right) \cong I_O \exp\left(\frac{I_{\mathrm{IN}}}{I_O}\right). \tag{1.34}$$

This particular form of g_{1c} function allows an important improvement of the output dynamic range of the exponential function synthesizer that is based on it.

1.3.2 *Second Class of Wide Output Dynamic Range Exponential Function Synthesizers with Second-Order Approximation*

1.3.2.1 Approximation Function Without Variable Changing

Another possible general form of a second-order approximation function is represented by

$$g_{2a}(x) = \frac{1}{1 + \mathrm{ax}} + \mathrm{bx} + c. \tag{1.35}$$

The condition of having an identity between the second-order series of $g_{2a}(x)$ and $f(x)$ functions imposes the particular values of *a*, *b*, and *c* parameters, resulting the following particular expression of $g_{2a}(x)$ function:

$$g_{2a}(x) = \frac{1}{1 - x\sqrt{p}} + (n - \sqrt{p})x + (m - 1). \tag{1.36}$$

This general approximation function can be particularized for obtaining the exponential function replacing the concrete values of *m*, *n*, and *p* parameters from the Taylor series expansion of the exponential function. It results:

$$g_{2a}(x) = \frac{1}{1 - \frac{x}{\sqrt{2}}} + \left(1 - \frac{1}{\sqrt{2}}\right)x. \tag{1.37}$$

The approximation error of $f(x) = \exp(x)$ function using $g_{2a}(x)$ function is approximately equal with the third-order error (for small values of x variable):

$$\varepsilon_{f(x)}^{g_{2a}(x)} \cong \frac{3\sqrt{2} - 2}{12} \frac{x^3}{\exp(x)}. \tag{1.38}$$

Table 1.4 Comparison between $g_{2a}(x)$ approximation function and $f(x) = \exp(x)$ function

x	$f(x)$ (dB)	$g_{2a}(x)$ (dB)	ε (dB)
−1	−8.69	−10.67	0.38
−0.8	−6.95	−7.87	0.06
−0.4	−3.48	−3.58	0.01
0.0	0.00	0.00	0.00
0.4	3.48	3.59	0.00
0.8	6.95	8.08	0.04
1	8.69	11.38	0.20

A comparison between $g_{2a}(x)$ approximation function and $f(x) = \exp(x)$ function is presented in Table 1.4. The graphical representations of $g_{2a}(x)$ and $f(x) = \exp(x)$ functions are shown in Fig. 1.15, while the graphical representation of the approximation error, $\varepsilon(x)$, defined as the difference between $g_{2a}(x)$ and $f(x) = \exp(x)$ functions, is presented in Fig. 1.16.

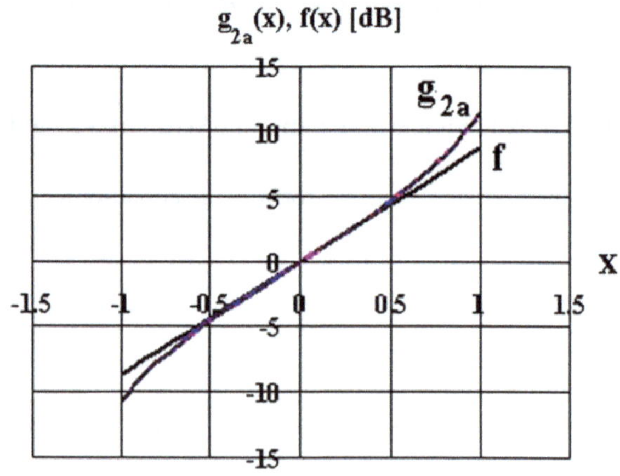

Fig. 1.15 Graphical representation of $g_{2a}(x)$ and $f(x) = \exp(x)$ functions

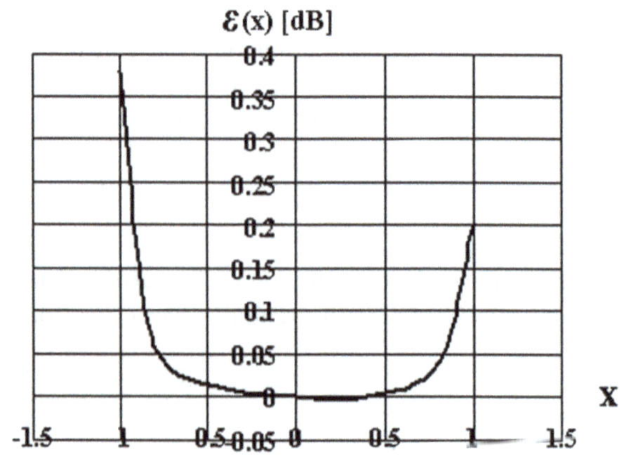

Fig. 1.16 Graphical representation of the approximation error, $\varepsilon(x)$

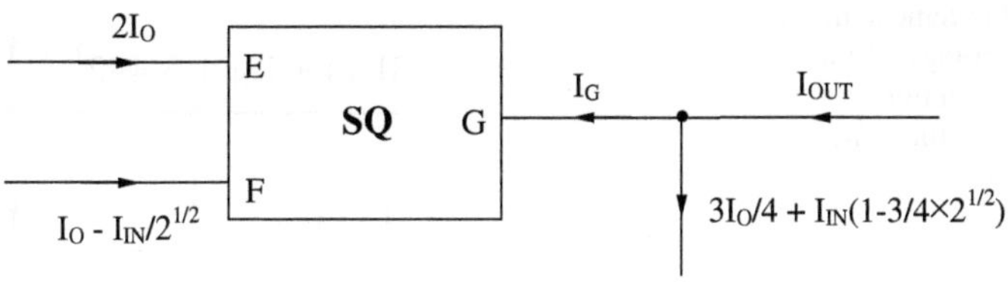

Fig. 1.17 Block diagram of the exponential function synthesizer circuit using $g_{2a}(x)$ approximation function

The block diagram of the exponential function synthesizer circuit using $g_{2a}(x)$ approximation function is shown in Fig. 1.17.

The I_G current can be expressed as follows:

$$I_G = \frac{1}{4}\left(I_O - \frac{I_{IN}}{\sqrt{2}}\right) - I_O + \frac{I_O^2}{I_O - \frac{I_{IN}}{\sqrt{2}}}. \tag{1.39}$$

So, it results the following expression of I_{OUT} current:

$$I_{OUT} = I_G + \left(1 - \frac{3}{4\sqrt{2}}\right) I_{IN} + \frac{3}{4} I_O = \frac{I_O^2}{I_O - \frac{I_{IN}}{\sqrt{2}}} + I_{IN}\left(1 - \frac{1}{\sqrt{2}}\right), \tag{1.40}$$

or

$$I_{OUT} = I_O\left[\frac{1}{1 - \frac{1}{\sqrt{2}}\left(\frac{I_{IN}}{I_O}\right)} + \left(1 - \frac{1}{\sqrt{2}}\right)\left(\frac{I_{IN}}{I_O}\right)\right]. \tag{1.41}$$

In conclusion, I_{OUT} current approximates the exponential function using $g_{2u}(x)$ approximation function:

$$I_{OUT} - I_O g_{2a}\left(\frac{I_{IN}}{I_O}\right) \cong I_O \exp\left(\frac{I_{IN}}{I_O}\right). \tag{1.42}$$

The implementation in CMOS technology of the exponential function synthesizer circuit that uses $g_{2a}(x)$ approximation function is shown in Fig. 1.18.

1.3.2.2 Approximation Function with First Variable Changing

The $x \to x/2$ variable can be used for increasing the maximal output dynamic range of the exponential function synthesizer based on $g_{2a}(x)$ approximation function. The modified approximation function can be expressed as follows:

$$g_{1b}(x) = \left[\frac{1}{1 - \frac{x}{2\sqrt{2}}} + \left(1 - \frac{1}{\sqrt{2}}\right)\frac{x}{2}\right]^2. \tag{1.43}$$

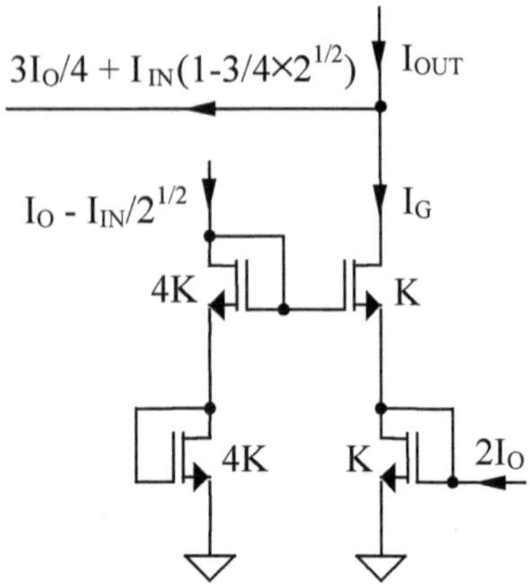

Fig. 1.18 Implementation in CMOS technology of the exponential function synthesizer circuit using $g_{2a}(x)$ approximation function

The comparison between $g_{2b}(x)$ approximation function and $f(x) = \exp(x)$ function is presented in Table 1.5. The graphical representations of $g_{2b}(x)$ and $f(x) = \exp(x)$ functions are shown in Fig. 1.19. In addition, the graphical representation of the approximation error, $\varepsilon(x)$, defined as the difference between $g_{2b}(x)$ and $f(x) = \exp(x)$ functions, is shown in Fig. 1.20.

The exponential function synthesizer' block diagram using $g_{2b}(x)$ approximation function is shown in Fig. 1.21.

The I_{Ga} current expression is:

$$I_{Ga} = \frac{1}{4}\left(I_O - \frac{I_{IN}}{2\sqrt{2}}\right) - I_O + \frac{I_O^2}{I_O - \frac{I_{IN}}{2\sqrt{2}}}, \tag{1.44}$$

resulting the following expression of $I_{OUT}{}'$ current:

$$I_{OUT}{}' = I_{Ga} + \left(\frac{1}{2} - \frac{3}{8\sqrt{2}}\right)I_{IN} + \frac{3}{4}I_O = \frac{I_O^2}{I_O - \frac{I_{IN}}{2\sqrt{2}}} + I_{IN}\left(\frac{1}{2} - \frac{1}{2\sqrt{2}}\right), \tag{1.45}$$

which is equivalent to:

Table 1.5 Comparison between $g_{2b}(x)$ approximation function and $f(x) = \exp(x)$ function

x	$f(x)$ (dB)	$g_{2b}(x)$ (dB)	ε (dB)
−1.6	−13.91	−15.73	−1.82
−1.2	−10.43	−11.15	−0.72
−0.8	−6.95	−7.16	−0.21
−0.4	−3.48	−3.50	−0.02
0.0	0.00	0.00	0.00
0.4	3.48	3.40	−0.08
0.8	6.95	7.18	0.23
1.2	10.43	11.27	0.84
1.6	13.91	16.17	2.26

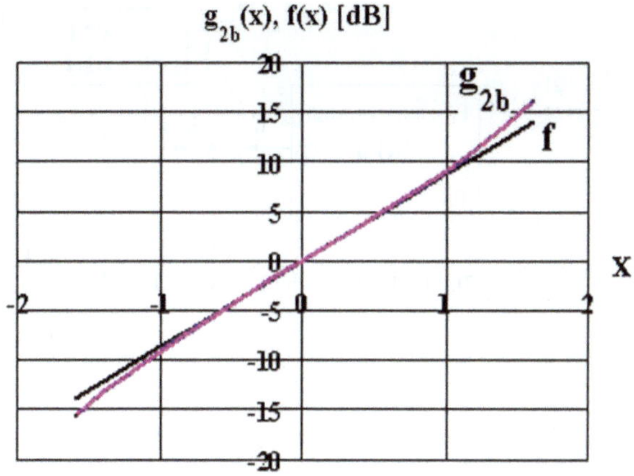

Fig. 1.19 Graphical representation of $g_{2b}(x)$ and $f(x) = \exp(x)$ functions

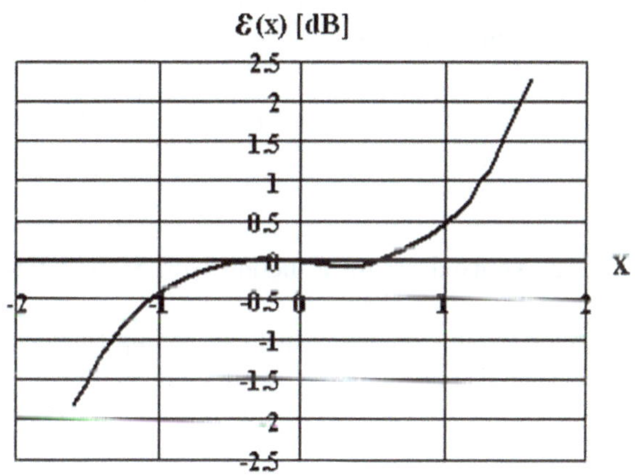

Fig. 1.20 Graphical representation of the approximation error, $\varepsilon(x)$

$$I_{OUT}' = I_O\left[\frac{1}{1-\frac{1}{2\sqrt{2}}\left(\frac{I_{IN}}{I_O}\right)} + \frac{1}{2}\left(1-\frac{1}{\sqrt{2}}\right)\left(\frac{I_{IN}}{I_O}\right)\right]. \tag{1.46}$$

The expression of I_{Gb} current is:

$$I_{Gb} = \frac{I_O}{16} - \frac{I_{OUT}'}{2} + \frac{(I_{OUT}')^2}{I_O}, \tag{1.47}$$

while I_{OUT} current can be expressed as follows:

$$I_{OUT} = I_{Gb} - \frac{I_O}{16} + \frac{I_{OUT}'}{2} = I_O\left[\frac{1}{1-\frac{1}{2\sqrt{2}}\left(\frac{I_{IN}}{I_O}\right)} + \frac{1}{2}\left(1-\frac{1}{\sqrt{2}}\right)\frac{I_{IN}}{I_O}\right]^2. \tag{1.48}$$

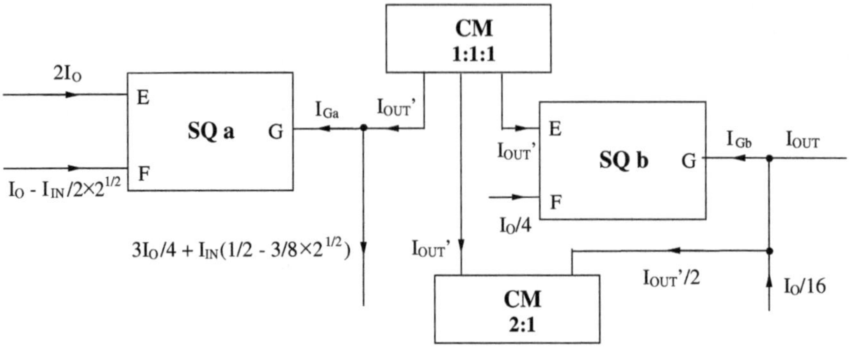

Fig. 1.21 Block diagram of the exponential function synthesizer circuit using $g_{2b}(x)$ approximation function

It results that I_{OUT} current approximates the exponential function using $g_{2b}(x)$ approximation function:

$$I_{\mathrm{OUT}} = I_O g_{2b}\left(\frac{I_{\mathrm{IN}}}{I_O}\right) \cong I_O \exp\left(\frac{I_{\mathrm{IN}}}{I_O}\right). \tag{1.49}$$

1.3.2.3 Approximation Function with Second Variable Changing

Another variable changing that can be implemented in CMOS technology with a reasonable increasing of the circuit complexity is $x \to x/4$. It results the following improved approximation function, having an increased output dynamic range:

$$g_{2c}(x) = \left[\frac{1}{1-\frac{x}{4\sqrt{2}}} + \left(1 - \frac{1}{\sqrt{2}}\right)\frac{x}{4}\right]^4. \tag{1.50}$$

A comparison between $g_{2c}(x)$ approximation function and $f(x) = \exp(x)$ function is presented in Table 1.6. The graphical representations of $g_{2c}(x)$ and $f(x) = \exp(x)$ functions are shown in Fig. 1.22, while the graphical representation of the approximation error, $\varepsilon(x)$, defined as the difference between $g_{2c}(x)$ and $f(x) = \exp(x)$ functions, is presented in Fig. 1.23.

The block diagram of the exponential function synthesizer circuit based on $g_{2c}(x)$ approximation function is presented in Fig. 1.24.

The expression of I_{Ga} current is:

$$I_{\mathrm{Ga}} = \frac{1}{4}\left(I_O - \frac{I_{\mathrm{IN}}}{4\sqrt{2}}\right) - I_O + \frac{I_O^2}{I_O - \frac{I_{\mathrm{IN}}}{4\sqrt{2}}}. \tag{1.51}$$

Table 1.6 Comparison between $g_{2c}(x)$ approximation function and $f(x) = \exp(x)$ function

x	$f(x)$ (dB)	$g_{2c}(x)$ (dB)	ε (dB)
−2.8	−24.34	−26.69	−2.35
−2.4	−20.86	−22.30	−1.44
−2.0	−17.38	−18.19	−0.81
−1.6	−13.91	−14.31	−0.40
−1.2	−10.43	−10.60	−0.17
−0.8	−6.95	−7.00	−0.05
−0.4	−3.48	−3.48	0.00
0.0	0.00	0.00	0.00
0.4	3.48	3.48	0.00
0.8	6.95	7.00	0.05
1.2	10.43	10.61	0.18
1.6	13.91	14.35	0.44
2.0	17.38	18.30	0.92
2.4	20.86	22.53	1.67
2.8	24.34	27.16	2.82

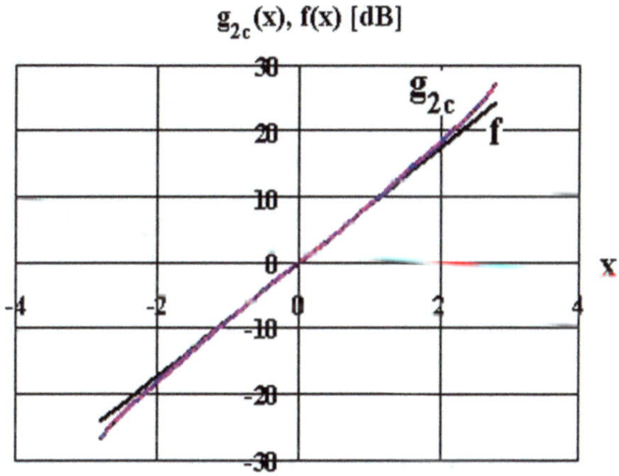

Fig. 1.22 Graphical representation of $g_{2c}(x)$ and $f(x) = \exp(x)$ functions

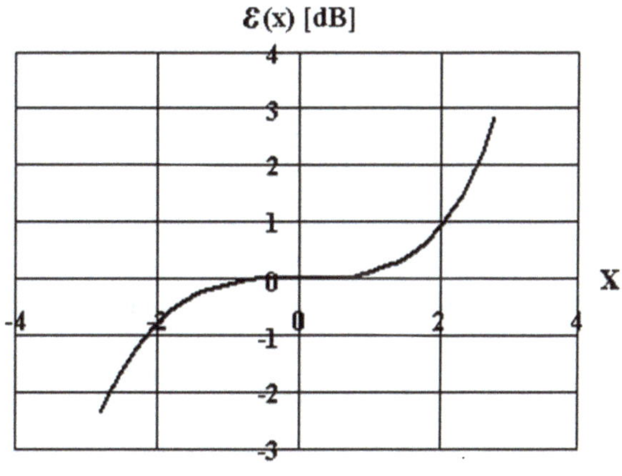

Fig. 1.23 Graphical representation of the approximation error, $\varepsilon(x)$

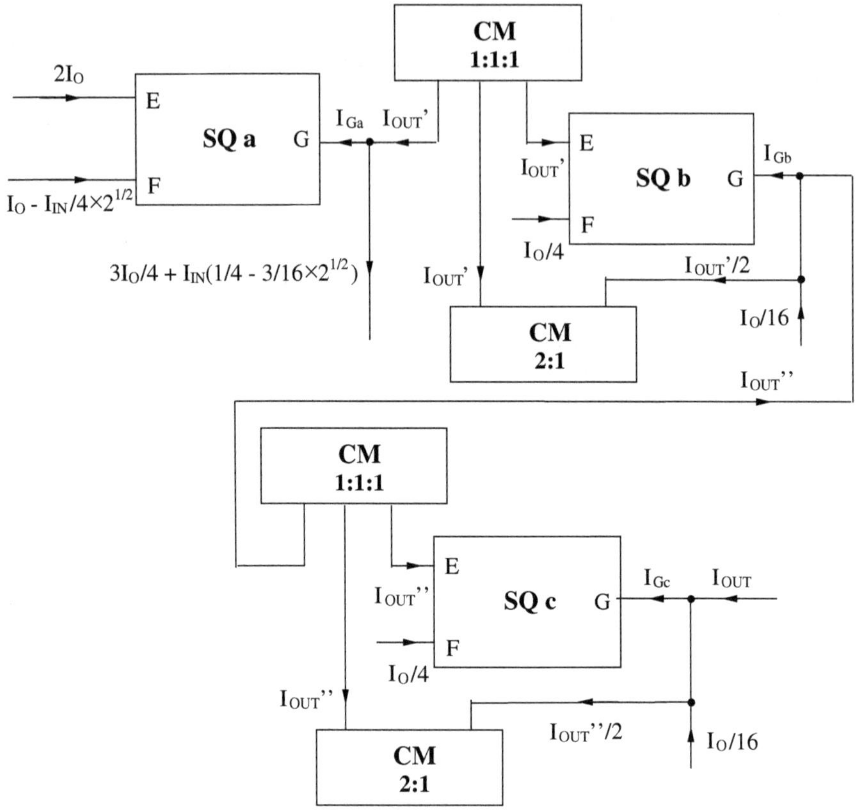

Fig. 1.24 Block diagram of the exponential function synthesizer circuit based on $g_{2c}(x)$ approximation function

It results the following expression of I_{OUT}' current:

$$I_{OUT}' = I_{G1} + \left(\frac{1}{4} - \frac{3}{16\sqrt{2}}\right)I_{IN} + \frac{3}{4}I_O = \frac{I_O^2}{I_O - \frac{I_{IN}}{4\sqrt{2}}} + I_{IN}\left(\frac{1}{4} - \frac{1}{4\sqrt{2}}\right). \tag{1.52}$$

So

$$I_{OUT}' = I_O\left[\frac{1}{1 - \frac{1}{4\sqrt{2}}\left(\frac{I_{IN}}{I_O}\right)} + \frac{1}{4}\left(1 - \frac{1}{\sqrt{2}}\right)\left(\frac{I_{IN}}{I_O}\right)\right]. \tag{1.53}$$

The I_{Gb} current will have the following expression:

$$I_{Gb} = \frac{I_O}{16} - \frac{I_{OUT}'}{2} + \frac{(I_{OUT}')^2}{I_O}, \tag{1.54}$$

while the expression of I_{OUT}'' current is:

$$I_{OUT}'' = I_{G2} - \frac{I_O}{16} + \frac{I_{OUT}'}{2} = I_O \left[\frac{1}{1 - \frac{1}{4\sqrt{2}} \left(\frac{I_{IN}}{I_O} \right)} + \frac{1}{4} \left(1 - \frac{1}{\sqrt{2}} \right) \left(\frac{I_{IN}}{I_O} \right) \right]^2. \quad (1.55)$$

The I_{Gc} current will be:

$$I_{Gc} = \frac{I_O}{16} - \frac{I_{OUT}''}{2} + \frac{(I_{OUT}'')^2}{I_O}, \quad (1.56)$$

resulting the following expression of I_{OUT} current:

$$I_{OUT} = I_{G3} - \frac{I_O}{16} + \frac{I_{OUT}''}{2} = I_O \left[\frac{1}{1 - \frac{1}{4\sqrt{2}} \left(\frac{I_{IN}}{I_O} \right)} + \frac{1}{4} \left(1 - \frac{1}{\sqrt{2}} \right) \left(\frac{I_{IN}}{I_O} \right) \right]^4. \quad (1.57)$$

Thus, I_{OUT} current approximates the exponential function using $g_{2c}(x)$ approximation function:

$$I_{OUT} = I_O g_{1c} \left(\frac{I_{IN}}{I_O} \right) \cong I_O \exp \left(\frac{I_{IN}}{I_O} \right). \quad (1.58)$$

The g_{2c} function contributes to an important improvement of the output dynamic range of the derived exponential function synthesizer.

1.4 Wide Output Dynamic Range Exponential Function Synthesizers Based on Third-Order Approximation Functions

1.4.1 First Class of Wide Output Dynamic Range Exponential Function Synthesizers with Third-Order Approximation

1.4.1.1 Approximation Function Without Variable Changing

The general form of an approximation function that can be relatively simple implemented in CMOS technology can be:

$$g_{3a}(x) = \frac{a}{1 + \text{bx}} + \text{cx} + d. \quad (1.59)$$

The particular values of *a*, *b*, *c*, and *d* coefficients can be determined from the condition of a third-order match between the Taylor series of $g_{3a}(x)$ and $f(x)$ functions, resulting:

$$g_{3a}(x) = \frac{p^3}{q^2}\frac{1}{1-\frac{qx}{p}} + \left(n - \frac{p^2}{q}\right)x + \left(m - \frac{p^3}{q^2}\right). \tag{1.60}$$

Replacing the coefficients from the Taylor series expansion associated with the exponential function, it results:

$$g_{3a}(x) = \frac{9}{2}\frac{1}{1-\frac{x}{3}} - \frac{x}{2} - \frac{7}{2}. \tag{1.61}$$

For small values of x variable, the approximation error for the exponential function using $g_{3a}(x)$ function can be approximated with the fourth-order error:

$$\varepsilon_{f(x)}^{g_{3a}(x)} \cong \frac{x^4}{72\exp(x)}. \tag{1.62}$$

The comparison between $g_{3a}(x)$ approximation function and $f(x) = \exp(x)$ function is presented in Table 1.7. The graphical representations of $g_{3a}(x)$ and $f(x) = \exp(x)$ functions are shown in Fig. 1.25, while the graphical representation of the approximation error, $\varepsilon(x)$, defined as the difference between $g_{3a}(x)$ and $f(x) = \exp(x)$ functions, is presented in Fig. 1.26.

The block diagram of the exponential function synthesizer circuit using $g_{3a}(x)$ approximation function is presented in Fig. 1.27.

The expression of I_G current is:

$$I_G = \frac{1}{4}\left(\frac{I_O}{2} - \frac{I_{IN}}{6}\right) - \frac{1}{2}3I_O + \frac{9I_O^2}{4\left(\frac{I_O}{2} - \frac{I_{IN}}{6}\right)}. \tag{1.63}$$

Table 1.7 Comparison between $g_{3a}(x)$ approximation function and $f(x) = \exp(x)$ function

x	$f(x)$ (dB)	$g_{3a}(x)$ (dB)	ε (dB)
−2.0	−17.38	−13.98	3.40
−1.6	−13.91	−12.59	1.32
−1.2	−10.43	−10.05	0.38
−0.8	−6.95	−6.89	0.06
−0.4	−3.48	−3.47	0.01
0.0	0.00	0.00	0.00
0.4	3.48	3.48	0.00
0.8	6.95	6.99	0.04
1.2	10.43	10.63	0.20
1.6	13.91	14.56	0.65
2.0	17.38	19.08	1.70

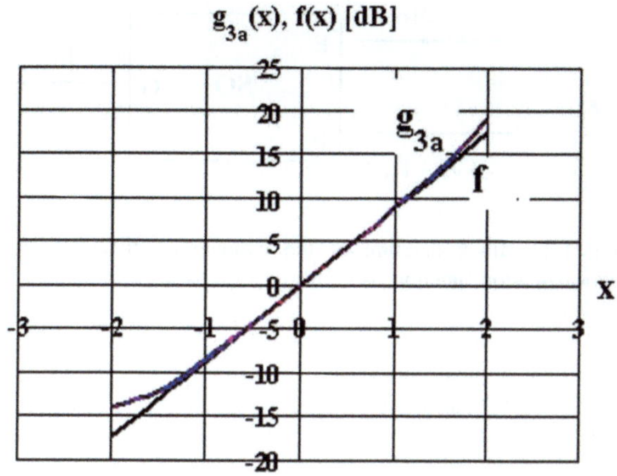

Fig. 1.25 Graphical representation of $g_{2a}(x)$ and $f(x) = \exp(x)$ functions

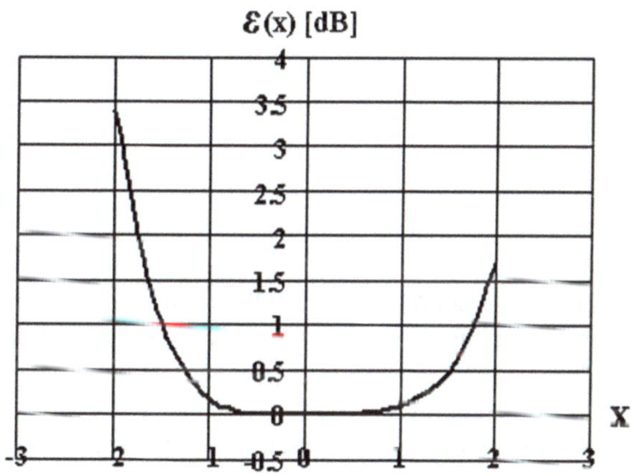

Fig. 1.26 Graphical representation of the approximation error, $\varepsilon(x)$

So, the I_{OUT} current can be expressed as follows:

$$I_{OUT} = I_G - \frac{11}{24} I_{IN} - \frac{17}{8} I_O = \frac{9}{2} \frac{I_O^2}{I_O - \frac{I_{IN}}{3}} - \frac{I_{IN}}{2} - \frac{7I_O}{2}, \tag{1.64}$$

which is equivalent to

$$I_{OUT} = I_O \left[\frac{9}{2} \frac{1}{1 - \frac{1}{3}\left(\frac{I_{IN}}{I_O}\right)} - \frac{1}{2}\left(\frac{I_{IN}}{I_O}\right) - \frac{7}{2} \right]. \tag{1.65}$$

Thus, I_{OUT} current approximates the exponential function using $g_{3a}(x)$ approximation function:

$$I_{OUT} = I_O g_{3a}\left(\frac{I_{IN}}{I_O}\right) \cong I_O \exp\left(\frac{I_{IN}}{I_O}\right). \tag{1.66}$$

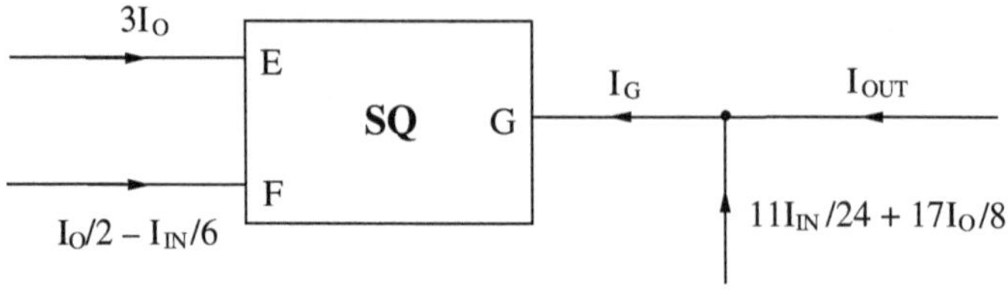

Fig. 1.27 Block diagram of the exponential function synthesizer circuit based on $g_{3a}(x)$ approximation function

Fig. 1.28 Implementation in CMOS technology of the exponential function synthesizer circuit using $g_{3a}(x)$ approximation function

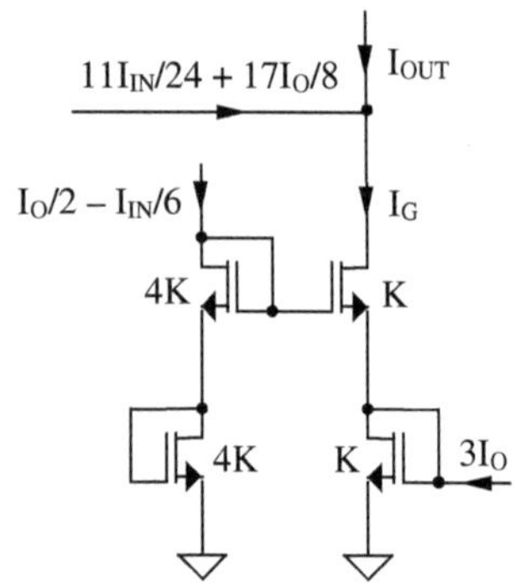

The implementation in CMOS technology of the exponential function synthesizer circuit based on $g_{3a}(x)$ approximation function is presented in Fig. 1.28.

1.4.1.2 Approximation Function with First Variable Changing

The maximal output dynamic range of the exponential function synthesizer based on $g_{3a}(x)$ approximation function can be increased using the $x \to x/2$ variable changing, resulting the following improved approximation function:

$$g_{3b}(x) = \left(\frac{9}{2}\frac{1}{1-\frac{x}{6}} - \frac{x}{4} - \frac{7}{2}\right)^2. \tag{1.67}$$

A comparison between $g_{3b}(x)$ approximation function and $f(x) = \exp(x)$ function is presented in Table 1.8. The graphical representations of $g_{3b}(x)$ and $f(x) = \exp(x)$ functions are shown in Fig. 1.29, while the graphical representation of the approximation error, $\varepsilon(x)$, defined as the difference between $g_{3b}(x)$ and $f(x) = \exp(x)$ functions, is presented in Fig. 1.30.

The block diagram of the exponential function synthesizer circuit based on $g_{3b}(x)$ approximation function is shown in Fig. 1.31.

The expression of I_{Ga} current is:

Table 1.8 Comparison between $g_{3b}(x)$ approximation function and $f(x) = \exp(x)$ function

x	$f(x)$ (dB)	$g_{3b}(x)$ (dB)	ε (dB)
−4.0	−34.75	−27.96	6.79
−3.6	−31.29	−26.91	4.38
−3.2	−27.81	−25.17	2.64
−2.8	−24.34	−22.86	1.48
−2.4	−20.86	−20.11	0.75
−2.0	−17.38	−17.04	0.34
−1.6	−13.91	−13.77	0.14
−1.2	−10.43	−10.39	0.04
−0.8	−6.95	−6.94	0.01
−0.4	−3.48	−3.47	0.01
0.0	0.00	0.00	0.00
0.4	3.48	3.47	0.01
0.8	6.95	6.96	0.01
1.2	10.43	10.45	0.02
1.6	13.91	13.98	0.07
2.0	17.38	17.57	0.19
2.4	20.86	21.26	0.40
2.8	24.34	25.08	0.72
3.2	27.81	29.11	1.30
3.6	31.29	33.43	2.14
4.0	34.75	38.16	3.41

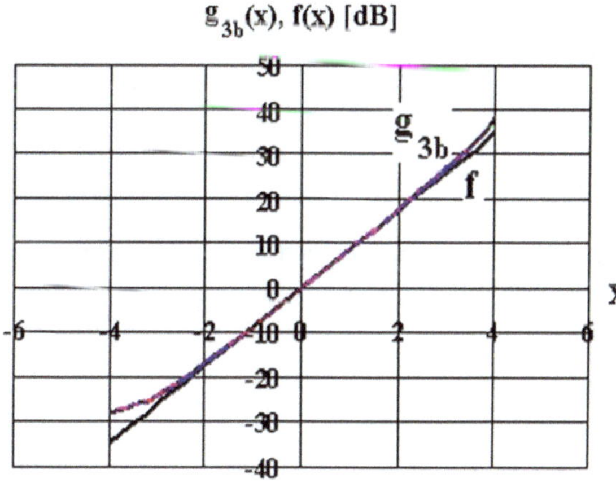

Fig. 1.29 Graphical representation of $g_{3b}(x)$ and $f(x) = \exp(x)$ functions

$$I_{Ga} = \frac{1}{4}\left(\frac{I_O}{2} - \frac{I_{IN}}{12}\right) - \frac{1}{2}3I_O + \frac{9I_O^2}{4\left(\frac{I_O}{2} - \frac{I_{IN}}{12}\right)}. \tag{1.68}$$

So, I_{OUT}' current will have the following expression:

$$I_{OUT}' = I_{Ga} - \frac{11}{48}I_{IN} - \frac{17}{8}I_O = \frac{9}{2}\frac{I_O^2}{I_O - \frac{I_{IN}}{6}} - \frac{I_{IN}}{4} - \frac{7I_O}{2}, \tag{1.69}$$

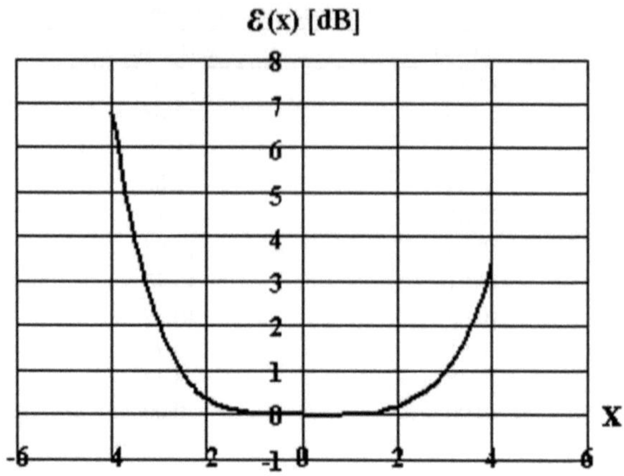

Fig. 1.30 Graphical representation of the approximation error, $\varepsilon(x)$

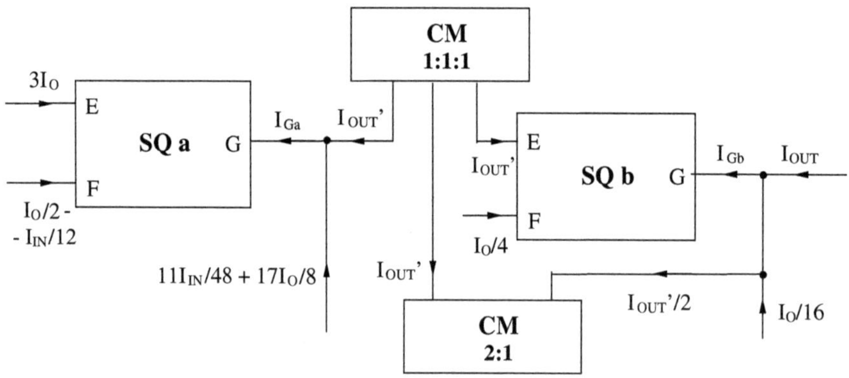

Fig. 1.31 Block diagram of the exponential function synthesizer circuit using $g_{3b}(x)$ approximation function

resulting

$$I_{\mathrm{OUT}}{}' = I_O \left[\frac{9}{2} \frac{1}{1 - \frac{1}{6}\left(\frac{I_{\mathrm{IN}}}{I_O}\right)} - \frac{1}{4}\left(\frac{I_{\mathrm{IN}}}{I_O}\right) - \frac{7}{2} \right]. \tag{1.70}$$

The I_{Gb} current can be expressed as follows:

$$I_{\mathrm{Gb}} = \frac{I_O}{16} - \frac{I_{\mathrm{OUT}}{}'}{2} + \frac{(I_{\mathrm{OUT}}{}')^2}{I_O}, \tag{1.71}$$

while the expression of I_{OUT}'' current is:

$$I_{OUT} = I_{Gb} - \frac{I_O}{16} + \frac{I_{OUT}'}{2} = I_O \left[\frac{9}{2} \frac{1}{1 - \frac{1}{6}\left(\frac{I_{IN}}{I_O}\right)} - \frac{1}{4}\left(\frac{I_{IN}}{I_O}\right) - \frac{7}{2} \right]^2. \tag{1.72}$$

In conclusion, I_{OUT} current approximates the exponential function using $g_{3b}(x)$ approximation function:

$$I_{OUT} = I_O g_{3a}\left(\frac{I_{IN}}{I_O}\right) \cong I_O \exp\left(\frac{I_{IN}}{I_O}\right). \tag{1.73}$$

1.4.1.3 Approximation Function with Second Variable Changing

In order to additionally increase the maximal output dynamic range of the exponential function synthesizer circuit, another variable changing can be used: $x \to x/4$. The improvement of the circuit performances is a consequence of the reduction in the domain of variable for the x input variable. The modified approximation function can be expressed as follows:

$$g_{1c}(x) = \left(\frac{9}{2} \frac{1}{1 - \frac{x}{12}} - \frac{x}{8} - \frac{7}{2} \right)^4. \tag{1.74}$$

A comparison between $g_{3c}(x)$ approximation function and $f(x) = \exp(x)$ function is presented in Table 1.9. The graphical representations of $g_{3c}(x)$ and $f(x) = \exp(x)$ functions are shown in Fig. 1.32, while the graphical representation of the approximation error, $\varepsilon(x)$, defined as the difference between $g_{3c}(x)$ and $f(x) = \exp(x)$ functions, is presented in Fig. 1.33.

The block diagram of the exponential function synthesizer circuit based on $g_{3c}(x)$ approximation function is presented in Fig. 1.34.

The expression of I_{Ga} current is

$$I_{Ga} = \frac{1}{4}\left(\frac{I_O}{2} - \frac{I_{IN}}{24}\right) - \frac{1}{2} 3 I_O + \frac{9 I_O^2}{4\left(\frac{I_O}{2} - \frac{I_{IN}}{24}\right)}. \tag{1.75}$$

So, I_{OUT}' current will have the following expression:

$$I_{OUT}' = I_{Ga} - \frac{11}{96} I_{IN} - \frac{17}{8} I_O = \frac{9}{2} \frac{I_O^2}{I_O - \frac{I_{IN}}{12}} - \frac{I_{IN}}{8} - \frac{7 I_O}{2}, \tag{1.76}$$

equivalent to

$$I_{OUT}' = I_O \left[\frac{9}{2} \frac{1}{1 - \frac{1}{12}\left(\frac{I_{IN}}{I_O}\right)} - \frac{1}{8}\left(\frac{I_{IN}}{I_O}\right) - \frac{7}{2} \right]. \tag{1.77}$$

Table 1.9 Comparison between $g_{3c}(x)$ approximation function and $f(x) = \exp(x)$ function

x	$f(x)$ (dB)	$g_{3c}(x)$ (dB)	ε (dB)
−5.2	−45.20	−43.06	2.14
−4.8	−41.72	−40.21	1.51
−4.4	−38.24	−37.21	1.03
−4.0	−34.75	−34.08	0.67
−3.6	−31.29	−30.85	0.44
−3.2	−27.81	−27.54	0.27
−2.8	−24.34	−24.18	0.16
−2.4	−20.86	−20.77	0.09
−2.0	−17.38	−17.34	0.04
−1.6	−13.91	−13.88	0.03
−1.2	−10.43	−10.42	0.01
−0.8	−6.95	−6.95	0.00
−0.4	−3.48	−3.48	0.00
0.0	0.00	0.00	0.00
0.4	3.48	3.48	0.00
0.8	6.95	6.95	0.00
1.2	10.43	10.43	0.00
1.6	13.91	13.91	0.00
2.0	17.38	17.40	0.02
2.4	20.86	20.90	0.04
2.8	24.34	24.42	0.08
3.2	27.81	27.96	0.15
3.6	31.29	31.54	0.25
4.0	34.75	35.15	0.40
4.4	38.24	38.80	0.56
4.8	41.72	42.52	0.80
5.2	45.20	46.30	1.10

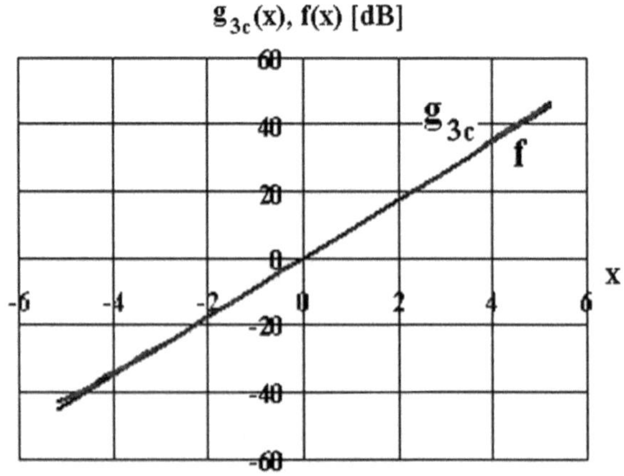

Fig. 1.32 Graphical representation of $g_{3c}(x)$ and $f(x) = \exp(x)$ functions

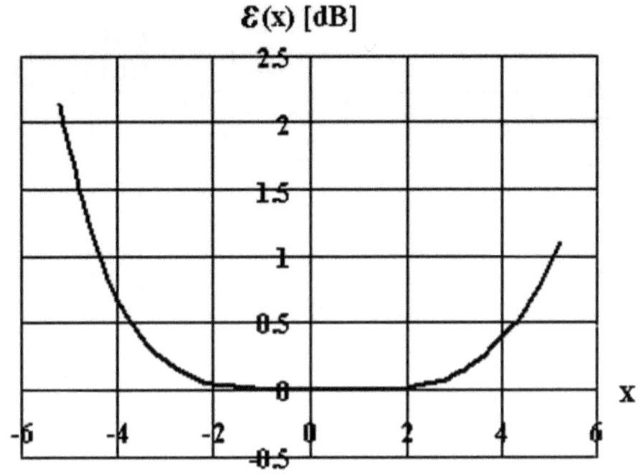

Fig. 1.33 Graphical representation of the approximation error, $\varepsilon(x)$

The I_{Gb} current can be expressed as follows:

$$I_{Gb} = \frac{I_O}{16} - \frac{I_{OUT}'}{2} + \frac{(I_{OUT}')^2}{I_O}, \tag{1.78}$$

while the expression of I_{OUT}'' current is:

$$I_{OUT}'' = I_{Gb} - \frac{I_O}{16} + \frac{I_{OUT}'}{2} = I_O\left[\frac{9}{2}\frac{1}{1 - \frac{1}{12}\left(\frac{I_{IN}}{I_O}\right)} - \frac{1}{8}\left(\frac{I_{IN}}{I_O}\right) - \frac{7}{2}\right]^2. \tag{1.79}$$

The expression of I_{Gc} current will be

$$I_{Gc} = \frac{I_O}{16} - \frac{I_{OUT}''}{2} + \frac{(I_{OUT}'')^2}{I_O}, \tag{1.80}$$

resulting the following expression of I_{OUT} current:

$$I_{OUT} = I_{Gc} - \frac{I_O}{16} + \frac{I_{OUT}''}{2} = I_O\left[\frac{9}{2}\frac{1}{1 - \frac{1}{12}\left(\frac{I_{IN}}{I_O}\right)} - \frac{1}{8}\left(\frac{I_{IN}}{I_O}\right) - \frac{7}{2}\right]^4. \tag{1.81}$$

So, I_{OUT} current approximates the exponential function using $g_{3c}(x)$ approximation function:

$$I_{OUT} = I_O g_{3c}\left(\frac{I_{IN}}{I_O}\right) \cong I_O \exp\left(\frac{I_{IN}}{I_O}\right). \tag{1.82}$$

This particular form of g_{1d} function allows an important improvement of the output dynamic range of the exponential function synthesizer that is based on it.

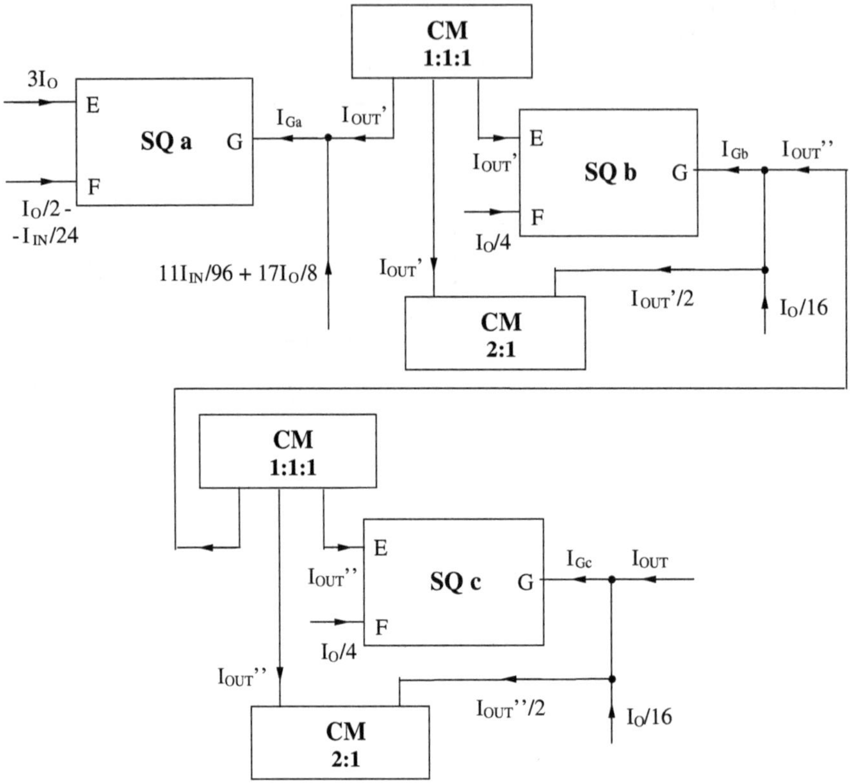

Fig. 1.34 Block diagram of the exponential function synthesizer circuit based on $g_{3c}(x)$ approximation function

1.4.2 Second Class of Wide Output Dynamic Range Exponential Function Synthesizers with Third-Order Approximation

1.4.2.1 Approximation Function Without Variable Changing

Another convenient form of a third-order approximation function, which can be used for generating the exponential function, has the following expression:

$$g_{4a}(x) = \frac{a}{1+x} + \mathrm{bx}^2 + \mathrm{cx} + d. \tag{1.83}$$

The complexity of the computational structure that implements this approximation function is relatively small, requiring only two current-mode squaring circuits. The concrete form of this function can be obtained determining the values of *a*, *b*, *c*, and *d* coefficients from the condition of a third-order match between the exponential and the approximation functions. It results:

$$g_{4a}(x) = -\frac{q}{1+x} + (p+q)x^2 + (n-q)x + (m+q). \tag{1.84}$$

The particular form of $g_{4a}(x)$ approximation function for generating the exponential function will be:

$$g_{4a}(x) = -\frac{1}{6}\frac{1}{1+x} + \frac{2x^2}{3} + \frac{5x}{6} + \frac{7}{6}. \tag{1.85}$$

If x variable has relatively small values, the approximation error for the $f(x) = \exp(x)$ function using $g_{4a}(x)$ function can be approximated with the fourth-order error. If the x input variable is increasing, more terms having an order greater than four must be considered. The error of approximation can be expressed as follows:

$$\varepsilon_{f(x)}^{g_{4a}(x)} \cong \frac{5x^4}{24\exp(x)}. \tag{1.86}$$

A comparison between $g_{4a}(x)$ approximation function and $f(x) = \exp(x)$ function is presented in Table 1.10. The graphical representations of $g_{4a}(x)$ and $f(x) = \exp(x)$ functions are shown in Fig. 1.35, while the graphical representation of the approximation error, $\varepsilon(x)$, defined as the difference between $g_{4a}(x)$ and $f(x) = \exp(x)$ functions, is presented in Fig. 1.36.

The block diagram of the exponential function synthesizer circuit based on $g_{4a}(x)$ approximation function is presented in Fig. 1.37.

The expression of I_{Ga} and I_{Gb} currents are

$$I_{Ga} = \frac{3}{8}(I_O + I_{IN}) - \frac{I_O}{2} + \frac{I_O^2}{6(I_O + I_{IN})} \tag{1.87}$$

and

$$I_{Gb} = \frac{3I_O}{32} - \frac{I_{IN}}{2} + \frac{2I_{IN}^2}{3I_O}, \tag{1.88}$$

Table 1.10 Comparison between $g_{4a}(x)$ approximation function and $f(x) = \exp(x)$ function

x	$f(x)$ (dB)	$g_{4a}(x)$ (dB)	ε (dB)
−0.6	−5.21	−6.20	0.99
−0.4	−3.48	−3.58	0.10
0.0	0.00	0.00	0.00
0.4	3.48	3.45	0.03
0.8	6.95	6.72	0.23
1.2	10.43	9.69	0.74
1.6	13.91	12.35	1.56
2.0	17.38	14.72	2.66

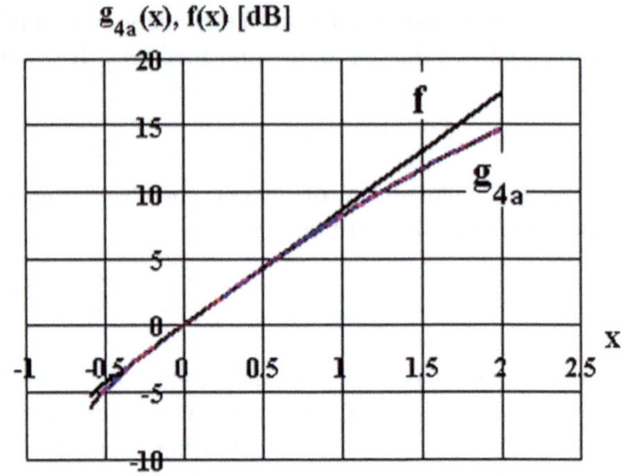

Fig. 1.35 Graphical representation of $g_{4a}(x)$ and $f(x) = \exp(x)$ functions

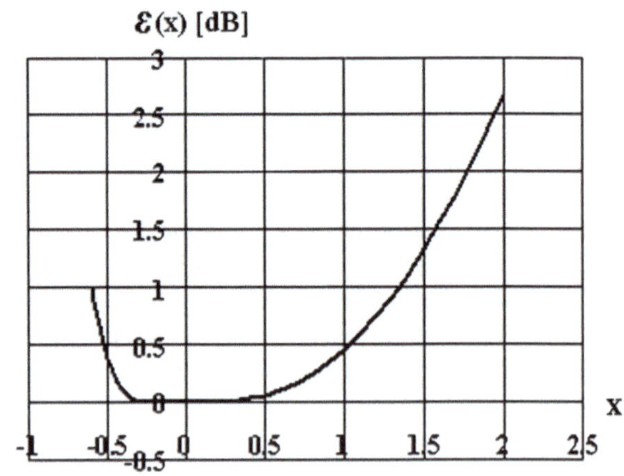

Fig. 1.36 Graphical representation of the approximation error, $\varepsilon(x)$

resulting the following expression of I_{OUT} current:

$$I_{OUT} = I_{Gb} - I_{Ga} + \frac{41}{24} I_{IN} + \frac{91}{96} I_O, \tag{1.89}$$

equivalent to

$$I_{OUT} = I_O \left[-\frac{1}{6} \frac{1}{1 + \left(\frac{I_{IN}}{I_O}\right)} + \frac{2}{3} \left(\frac{I_{IN}}{I_O}\right)^2 + \frac{5}{6} \left(\frac{I_{IN}}{I_O}\right) + \frac{7}{6} \right]. \tag{1.90}$$

So, I_{OUT} current approximates the exponential function using $g_{4a}(x)$ approximation function:

$$I_{OUT} = I_O g_{4a}\left(\frac{I_{IN}}{I_O}\right) \cong I_O \exp\left(\frac{I_{IN}}{I_O}\right). \tag{1.91}$$

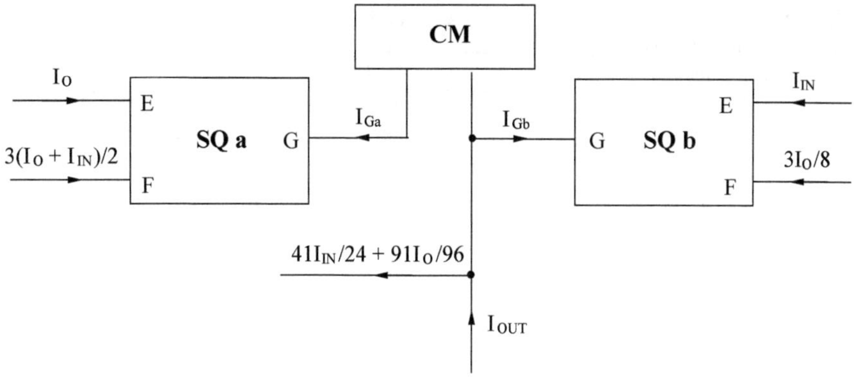

Fig. 1.37 Block diagram of the exponential function synthesizer circuit based on $g_{4a}(x)$ approximation function

1.4.2.2 Approximation Function with First Variable Changing

The increasing of the output dynamic range of the exponential function synthesizer circuit based on $g_{4b}(x)$ approximation function can be obtained considering the $x \to x/2$ variable changing. It results the following improved performances approximation function:

$$g_{4b}(x) = \left(\frac{1}{6} \frac{1}{1 + \frac{x}{2}} + \frac{x^2}{6} + \frac{5x}{12} + \frac{7}{6} \right)^2. \tag{1.92}$$

A comparison between $g_{4b}(x)$ approximation function and $f(x) = \exp(x)$ function is presented in Table 1.11. The graphical representations of $g_{4b}(x)$ and $f(x) = \exp(x)$ functions are shown in Fig. 1.38, while the graphical representation of the approximation error, $\varepsilon(x)$, defined as the difference between $g_{4b}(x)$ and $f(x) = \exp(x)$ functions, is presented in Fig. 1.39.

Table 1.11 Comparison between $g_{4b}(x)$ approximation function and $f(x) = \exp(x)$ function

x	$f(x)$ (dB)	$g_{4b}(x)$ (dB)	ε (dB)
−1.2	−10.43	−12.39	1.96
−0.8	−6.95	−7.16	0.21
−0.4	−3.48	−3.48	0.00
0.0	0.00	0.00	0.00
0.4	3.48	3.47	0.01
0.8	6.95	6.90	0.05
1.2	10.43	10.24	0.19
1.6	13.91	13.44	0.47
2.0	17.38	16.49	0.89
2.4	20.86	19.38	1.48
2.8	24.34	22.11	2.23
3.2	27.81	24.69	3.12

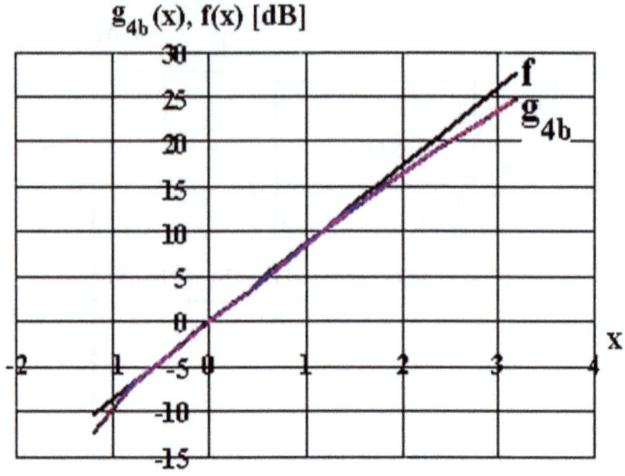

Fig. 1.38 Graphical representation of $g_{4b}(x)$ and $f(x) = \exp(x)$ functions

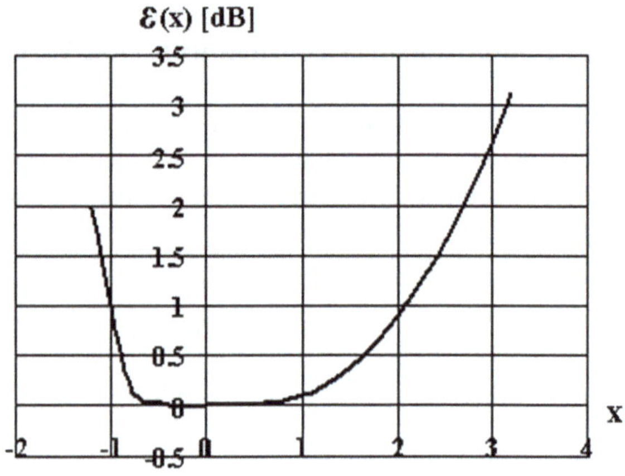

Fig. 1.39 Graphical representation of the approximation error, $\varepsilon(x)$

The block diagram of the exponential function synthesizer circuit based on $g_{4b}(x)$ approximation function is presented in Fig. 1.40.

The I_{Ga} and I_{Gb} currents can be expressed as follows:

$$I_{Ga} = \frac{1}{4}\left(\frac{3I_O}{2} + \frac{3I_{IN}}{4}\right) - \frac{I_O}{2} + \frac{I_O^2}{4\left(\frac{3I_O}{2} + \frac{3I_{IN}}{4}\right)} \tag{1.93}$$

and

$$I_{Gb} = \frac{3I_O}{8} - \frac{I_{IN}}{2} + \frac{I_{IN}^2}{6I_O}. \tag{1.94}$$

So, I_{OUT}' current will have the following expression:

$$I_{OUT}' = I_{Gb} - I_{Ga} + \frac{53}{48}I_{IN} + \frac{2}{3}I_O, \tag{1.95}$$

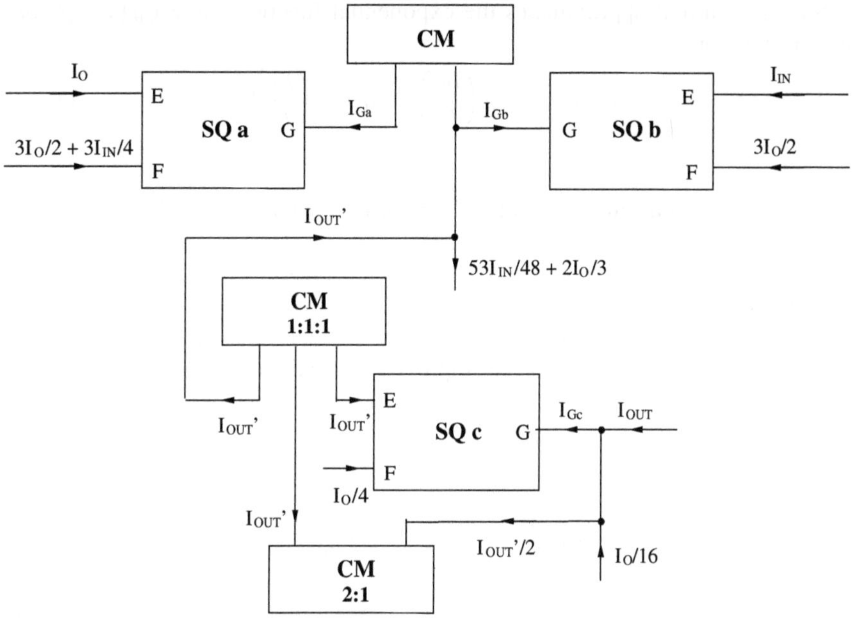

Fig. 1.40 Block diagram of the exponential function synthesizer circuit based on $g_{4b}(x)$ approximation function

equivalent to

$$I_{OUT}' = I_O\left[-\frac{1}{6}\frac{1}{1+\frac{1}{2}\left(\frac{I_{IN}}{I_O}\right)} + \frac{1}{6}\left(\frac{I_{IN}}{I_O}\right)^2 + \frac{5}{12}\left(\frac{I_{IN}}{I_O}\right) + \frac{7}{6}\right]. \tag{1.96}$$

The I_{Gc} current will have the following expression:

$$I_{Gc} = \frac{I_O}{16} - \frac{I_{OUT}'}{2} + \frac{(I_{OUT}')^2}{I_O}. \tag{1.97}$$

The expression of I_{OUT} current is

$$I_{OUT} = I_{Gc} - \frac{I_O}{16} + \frac{I_{OUT}'}{2}, \tag{1.98}$$

which is equivalent to

$$I_{OUT} = I_O\left[-\frac{1}{6}\frac{1}{1+\frac{1}{2}\left(\frac{I_{IN}}{I_O}\right)} + \frac{1}{6}\left(\frac{I_{IN}}{I_O}\right)^2 + \frac{5}{12}\left(\frac{I_{IN}}{I_O}\right) + \frac{7}{6}\right]^2. \tag{1.99}$$

So, I_{OUT} current approximates the exponential function using $g_{4b}(x)$ approximation function:

$$I_{OUT} = I_O g_{4b}\left(\frac{I_{IN}}{I_O}\right) \cong I_O \exp\left(\frac{I_{IN}}{I_O}\right). \tag{1.100}$$

1.4.2.3 Approximation Function with Second Variable Changing

The second variable changing that can be implemented in CMOS technology with a small additional increasing of the circuit complexity, but which strongly increases the output dynamic range of the exponential function synthesizer circuit is $x \to x/4$, resulting the following improved approximation function:

$$g_{2c}(x) = \left(-\frac{1}{6}\frac{1}{1+\frac{x}{4}} + \frac{x^2}{24} + \frac{5x}{24} + \frac{7}{6}\right)^4. \tag{1.101}$$

A comparison between $g_{4c}(x)$ approximation function and $f(x) = \exp(x)$ function is presented in Table 1.12. The graphical representations of $g_{4c}(x)$ and $f(x) = \exp(x)$ functions are shown in Fig. 1.41, while the graphical representation of the approximation error, $\varepsilon(x)$, defined as the difference between $g_{4c}(x)$ and $f(x) = \exp(x)$ functions, is presented in Fig. 1.42.

The block diagram of the exponential function synthesizer circuit based on $g_{4c}(x)$ approximation function is presented in Fig. 1.43.

The expression of I_{Ga} and I_{Gb} currents are

$$I_{Ga} = \frac{1}{4}\left(\frac{3I_O}{2} + \frac{3I_{IN}}{8}\right) - \frac{I_O}{2} + \frac{I_O^2}{4\left(\frac{3I_O}{2} + \frac{3I_{IN}}{8}\right)} \tag{1.102}$$

Table 1.12 Comparison between $g_{4c}(x)$ approximation function and $f(x) = \exp(x)$ function

x	$f(x)$ (dB)	$g_{4c}(x)$ (dB)	ε (dB)
−2.0	−17.38	−18.73	−1.35
−1.6	−13.91	−14.32	−0.41
−1.2	−10.43	−10.53	−0.10
−0.8	−6.95	−6.96	−0.01
−0.4	−3.48	−3.48	0.00
0.0	0.00	0.00	0.00
0.4	3.48	3.47	−0.01
0.8	6.95	6,94	−0.01
1.2	10.43	10,39	−0.04
1.6	13.91	13.80	−0.11
2.0	17.38	17.16	−0.22
2.4	20.86	20.47	−0.39
2.8	24.34	23.71	−0.63
3.2	27.81	26.88	−0.93
3.6	31.29	29.96	−1.33
4.0	34.75	32.97	−1.78

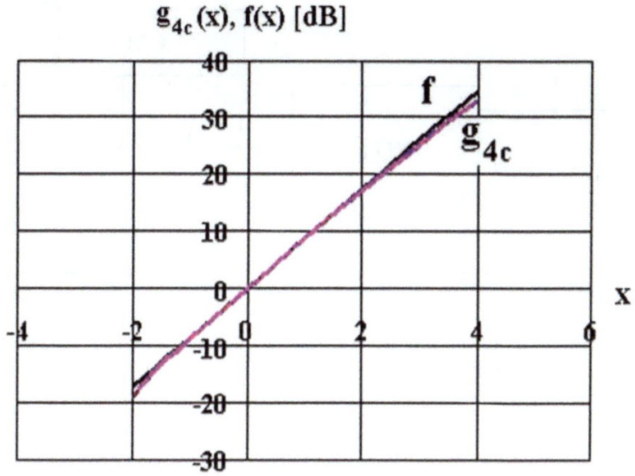

Fig. 1.41 Graphical representation of $g_{4c}(x)$ and $f(x) = \exp(x)$ functions

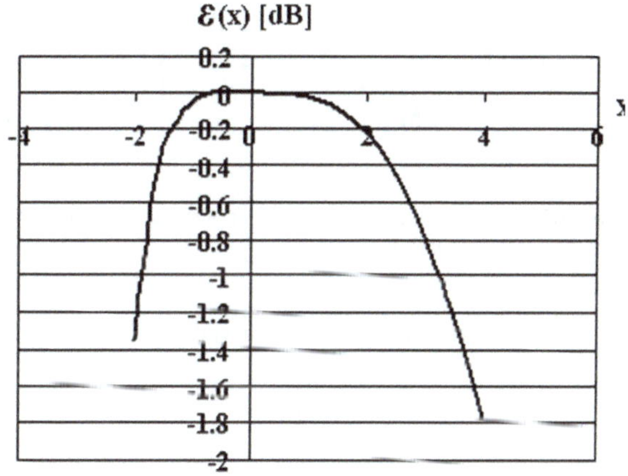

Fig. 1.42 Graphical representation of the approximation error, $\varepsilon(x)$

and

$$I_{G2} = \frac{3I_O}{2} - \frac{I_{IN}}{2} + \frac{I_{IN}^2}{24I_O}. \tag{1.103}$$

So, I_{OUT}' current will have the following expression:

$$I_{OUT}' = I_{Gb} - I_{Ga} + \frac{77}{96}I_{IN} - \frac{11}{24}I_O, \tag{1.104}$$

resulting

$$I_{OUT}' = I_O\left[-\frac{1}{6}\frac{1}{1+\frac{1}{4}\left(\frac{I_{IN}}{I_O}\right)} + \frac{1}{24}\left(\frac{I_{IN}}{I_O}\right)^2 + \frac{5}{24}\left(\frac{I_{IN}}{I_O}\right) + \frac{7}{6}\right]. \tag{1.105}$$

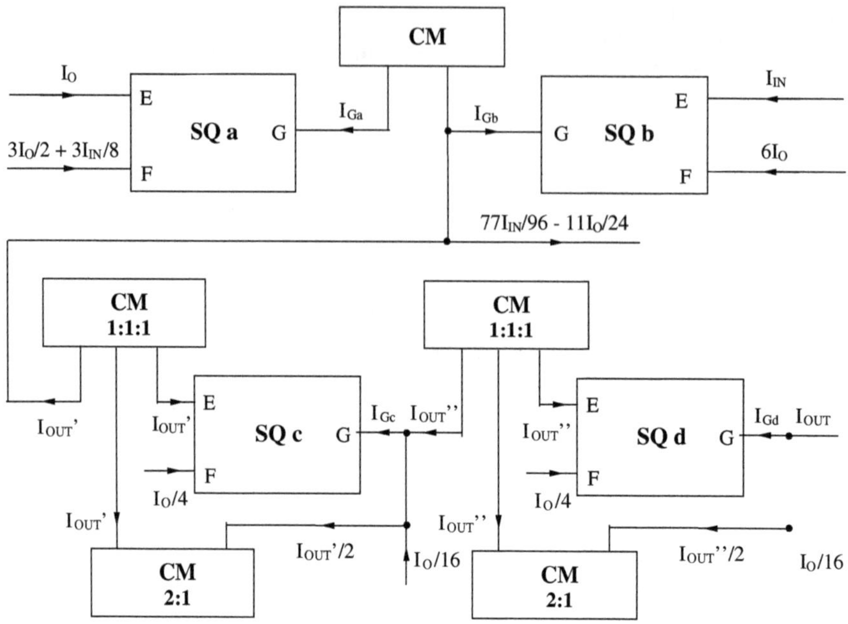

Fig. 1.43 Block diagram of the exponential function synthesizer circuit based on $g_{4c}(x)$ approximation function

The I_{Gc} current can be expressed as follows:

$$I_{Gc} = \frac{I_O}{16} - \frac{I_{OUT}'}{2} + \frac{(I_{OUT}')^2}{I_O}, \tag{1.106}$$

while the expression of I_{OUT}'' current is

$$I_{OUT}'' = I_{Gc} - \frac{I_O}{16} + \frac{I_{OUT}'}{2}, \tag{1.107}$$

equivalent to

$$I_{OUT}'' = I_O\left[-\frac{1}{6}\frac{1}{1+\frac{1}{4}\left(\frac{I_{IN}}{I_O}\right)} + \frac{1}{24}\left(\frac{I_{IN}}{I_O}\right)^2 + \frac{5}{24}\left(\frac{I_{IN}}{I_O}\right) + \frac{7}{6}\right]^2. \tag{1.108}$$

The expression of I_{Gd} current will be

$$I_{Gd} = \frac{I_O}{16} - \frac{I_{OUT}''}{2} + \frac{(I_{OUT}'')^2}{I_O}, \tag{1.109}$$

resulting the following expression of I_{OUT} current:

$$I_{OUT} = I_{Gd} - \frac{I_O}{16} + \frac{I_{OUT}''}{2}, \tag{1.110}$$

equivalent to

$$I_{OUT} = I_O \left[-\frac{1}{6} \frac{1}{1 + \frac{1}{4}\left(\frac{I_{IN}}{I_O}\right)} + \frac{1}{24}\left(\frac{I_{IN}}{I_O}\right)^2 + \frac{5}{24}\left(\frac{I_{IN}}{I_O}\right) + \frac{7}{6} \right]^4. \tag{1.111}$$

In conclusion, I_{OUT} current approximates the exponential function using $g_{4c}(x)$ approximation function:

$$I_{OUT} = I_O g_{4c}\left(\frac{I_{IN}}{I_O}\right) \cong I_O \exp\left(\frac{I_{IN}}{I_O}\right). \tag{1.112}$$

1.5 Wide Output Dynamic Range Function Synthesizers Based on Fourth-Order Approximation Functions

For additionally improving the approximation accuracy, the order of the approximation must be increased. The fourth order approximation functions present extremely low errors, requiring reasonable complexity of computational structures that use them.

1.5.1 First Class of Wide Output Dynamic Range Exponential Function Synthesizers with Fourth-Order Approximation

1.5.1.1 Approximation Function Without Variable Changing

The first presented fourth-order approximation function uses a fractional and a squaring term, their CMOS implementation requiring only two current-mode squaring circuits:

$$g_{5a}(x) = \frac{a}{1 + \mathrm{bx}} + \mathrm{cx}^2 + \mathrm{dx} + e. \tag{1.113}$$

The particular form of $g_{5a}(x)$ approximation function, derived from (1.113) and developed for approximating the exponential function, can be obtained from the fourth-order identity between the Taylor series of $f(x) = \exp(x)$ and $g_{5a}(x)$ functions. It results

$$g_{5a}(x) = \frac{32}{3}\frac{1}{1-\frac{x}{4}} - \frac{x^2}{4} - \frac{5x}{3} - \frac{29}{3}. \quad (1.114)$$

For small values of x variable, the approximation error for the $f(x) = \exp(x)$ function using $g_{5a}(x)$ function can be approximated with the fifth-order error:

$$\varepsilon_{f(x)}^{g_{5a}(x)} \cong \frac{x^5}{96\exp(x)}. \quad (1.115)$$

If the x input variable is increasing, more terms having an order greater than five must be considered. A comparison between $g_{5a}(x)$ approximation function and $f(x) = \exp(x)$ function is presented in Table 1.13. The graphical representations of $g_{5a}(x)$ and $f(x) = \exp(x)$ functions are shown in Fig. 1.44, while the graphical representation of the approximation error, $\varepsilon(x)$, defined as the difference between $g_{5a}(x)$ and $f(x) = \exp(x)$ functions, is presented in Fig. 1.45.

The block diagram of the exponential function synthesizer circuit based on $g_{5a}(x)$ approximation function is presented in Fig. 1.46.

Table 1.13 Comparison between $g_{5a}(x)$ approximation function and $f(x) = \exp(x)$ function

x	$f(x)$ (dB)	$g_{5a}(x)$ (dB)	ε (dB)
−1.0	−8.69	−10.95	2.26
−0.8	−6.95	−8.06	1.11
−0.4	−3.48	−3.65	0.17
0.0	0.00	0.00	0.00
0.4	3.48	3.40	0.08
0.8	6.95	6.74	0.21
1.2	10.43	10.13	0.30
1.6	13.91	13.63	0.28
2.0	17.38	17.31	0.07
2.4	20.86	21.26	−0.40
2.8	24.34	25.69	−1.35

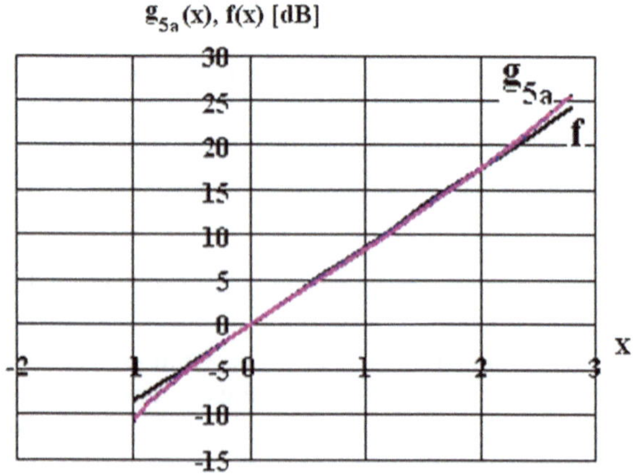

Fig. 1.44 Graphical representation of $g_{5a}(x)$ and $f(x) = \exp(x)$ functions

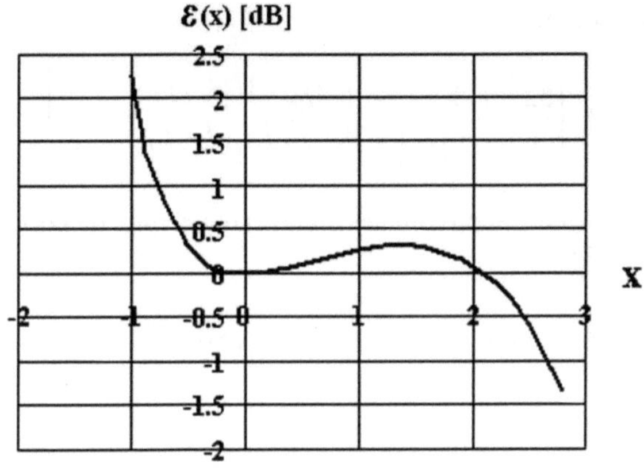

Fig. 1.45 Graphical representation of the approximation error, $\varepsilon(x)$

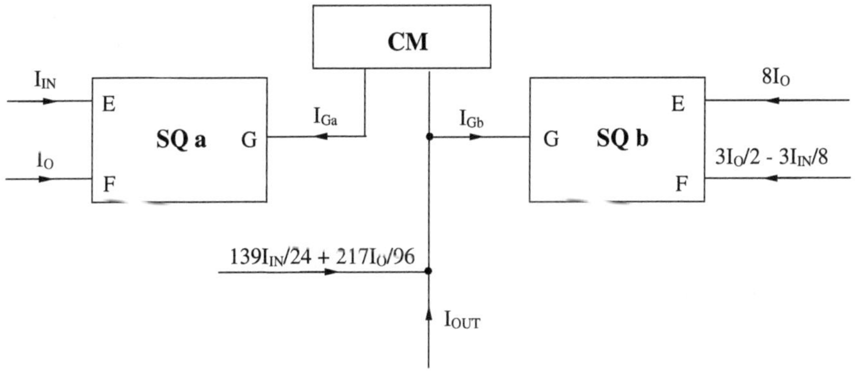

Fig. 1.46 Block diagram of the exponential function synthesizer circuit based on $g_{5a}(x)$ approximation function

The I_{Ga} and I_{Gb} currents can be expressed as follows:

$$I_{\mathrm{Ga}} = \frac{I_O}{4} - \frac{I_{\mathrm{IN}}}{2} + \frac{I_{\mathrm{IN}}^2}{4I_O} \tag{1.116}$$

and

$$I_{\mathrm{Gb}} = -\frac{3}{32}I_{\mathrm{IN}} + \frac{3}{8}I_O - 4I_O + \frac{32I_O^2}{3I_O - \frac{3}{4}I_{\mathrm{IN}}}, \tag{1.117}$$

resulting the following expression of I_{OUT} current:

$$I_{\mathrm{OUT}} = I_{\mathrm{Gb}} - I_{\mathrm{Ga}} - \frac{139}{24}I_O - \frac{199}{96}I_{\mathrm{IN}}. \tag{1.118}$$

This is equivalent to

$$I_{OUT} = I_O\left[\frac{32}{3}\frac{1}{1-\frac{1}{4}\left(\frac{I_{IN}}{I_O}\right)} - \frac{1}{4}\left(\frac{I_{IN}}{I_O}\right)^2 - \frac{5}{3}\left(\frac{I_{IN}}{I_O}\right) - \frac{29}{3}\right] \quad (1.119)$$

Thus, I_{OUT} current approximates the exponential function using $g_{5a}(x)$ approximation function:

$$I_{OUT} = I_O g_{5a}\left(\frac{I_{IN}}{I_O}\right) \cong I_O \exp\left(\frac{I_{IN}}{I_O}\right). \quad (1.120)$$

1.5.1.2 Approximation Function with First Variable Changing

In order to increase the maximal output dynamic range of the exponential function synthesizer based on $g_{5a}(x)$ approximation function, the $x \to x/2$ variable changing can be used. This variable changing can be implemented in CMOS technology with a reasonable increasing of the circuit complexity, resulting the following improved approximation function:

$$g_{5b}(x) = \left(\frac{32}{3}\frac{1}{1-\frac{x}{8}} - \frac{x^2}{16} - \frac{5x}{6} - \frac{29}{3}\right)^2. \quad (1.121)$$

The comparison between $g_{5b}(x)$ approximation function and $f(x) = \exp(x)$ function is presented in Table 1.14. The graphical representations of $g_{5b}(x)$ and $f(x) = \exp(x)$ functions are shown in Fig. 1.47, while the graphical representation of the approximation error, $\varepsilon(x)$, defined as the difference between $g_{5b}(x)$ and $f(x) = \exp(x)$ functions, is presented in Fig. 1.48.

The block diagram of the exponential function synthesizer circuit based on $g_{5b}(x)$ approximation function is presented in Fig. 1.49.

The expression of I_{Ga} and I_{Gb} currents are

$$I_{Ga} = I_O - \frac{I_{IN}}{2} + \frac{I_{IN}^2}{16 I_O}, \quad (1.122)$$

while

$$I_{Gb} = \frac{1}{4}\left(\frac{3I_O}{2} - \frac{3I_{IN}}{16}\right) - 4I_O + \frac{64 I_O^2}{4\left(\frac{3I_O}{2} - \frac{3I_{IN}}{16}\right)}. \quad (1.123)$$

So, I_{OUT}' current will have the following expression:

$$I_{OUT}' = I_{Gb} - I_{Ga} - \frac{247}{192} I_{IN} - \frac{121}{24} I_O, \quad (1.124)$$

equivalent to

Table 1.14 Comparison between $g_{5b}(x)$ approximation function and $f(x) = \exp(x)$ function

x	$f(x)$ (dB)	$g_{5b}(x)$ (dB)	ε (dB)
−1.6	−13.91	−16.11	2.20
−1.2	−10.43	−11.40	0.97
−0.8	−6.95	−7.30	0.35
−0.4	−3.48	−3.55	0.07
0.0	0.00	0.00	0.00
0.4	3.48	3.43	0.05
0.8	6.95	6.79	0.16
1.2	10.43	10.14	0.29
1.6	13.91	13.49	0.42
2.0	17.38	16.86	0.52
2.4	20.86	20.27	0.59
2.8	24.34	23.73	0.61
3.2	27.81	27.27	0.54
3.6	31.29	30.88	0.41
4.0	34.75	34.61	0.14
4.4	38.24	38.48	−0.24
4.8	41.72	42.52	−0.80
5.2	45.20	46.79	−1.59
5.6	48.67	51.39	−2.72

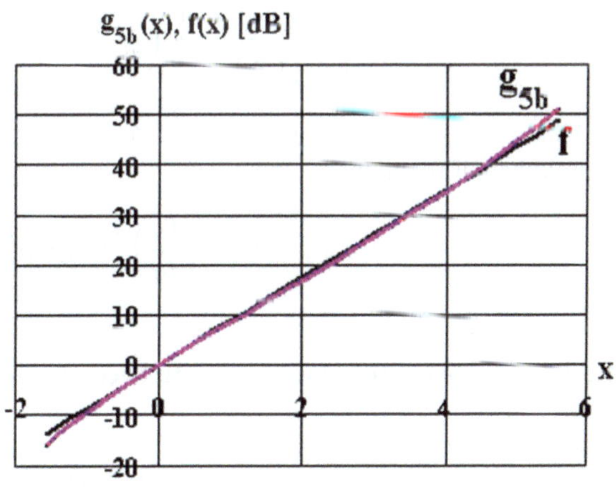

Fig. 1.47 Graphical representation of $g_{5b}(x)$ and $f(x) = \exp(x)$ functions

$$I_{OUT}' = I_O\left[\frac{32}{3}\frac{1}{1-\frac{1}{8}\left(\frac{I_{IN}}{I_O}\right)} - \frac{1}{16}\left(\frac{I_{IN}}{I_O}\right)^2 - \frac{5}{6}\left(\frac{I_{IN}}{I_O}\right) - \frac{29}{3}\right]. \tag{1.125}$$

The I_{Gc} current will have the following expression:

$$I_{Gc} = \frac{I_O}{16} - \frac{I_{OUT}'}{2} + \frac{(I_{OUT}')^2}{I_O}, \tag{1.126}$$

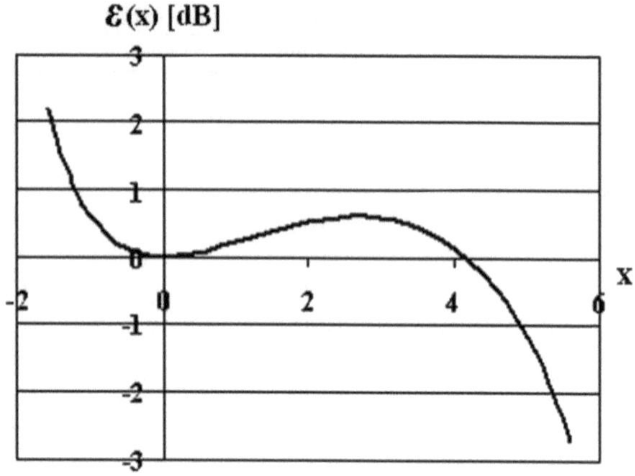

Fig. 1.48 Graphical representation of the approximation error, $\varepsilon(x)$

while the expression of I_{OUT} current is

$$I_{OUT} = I_{Gc} - \frac{I_O}{16} + \frac{I_{OUT}'}{2}, \tag{1.127}$$

equivalent to

$$I_{OUT} = I_O \left[\frac{32}{3} \frac{1}{1 - \frac{1}{8}\left(\frac{I_{IN}}{I_O}\right)} - \frac{1}{16}\left(\frac{I_{IN}}{I_O}\right)^2 - \frac{5}{6}\left(\frac{I_{IN}}{I_O}\right) - \frac{29}{3} \right]^2. \tag{1.128}$$

Thus, I_{OUT} current approximates the exponential function using $g_{5b}(x)$ approximation function:

$$I_{OUT} = I_O g_{5b}\left(\frac{I_{IN}}{I_O}\right) \cong I_O \exp\left(\frac{I_{IN}}{I_O}\right). \tag{1.129}$$

1.5.1.3 Approximation Function with Second Variable Changing

The second variable changing that can be implemented in CMOS technology with a small additional increasing of the circuit complexity, but which strongly increase the output dynamic range of the exponential function synthesizer circuit is $x \to x/4$, resulting the following improved approximation function:

$$g_{5c}(x) = \left(\frac{32}{3} \frac{1}{1 - \frac{x}{16}} - \frac{x^2}{64} - \frac{5x}{12} - \frac{29}{3} \right)^4. \tag{1.130}$$

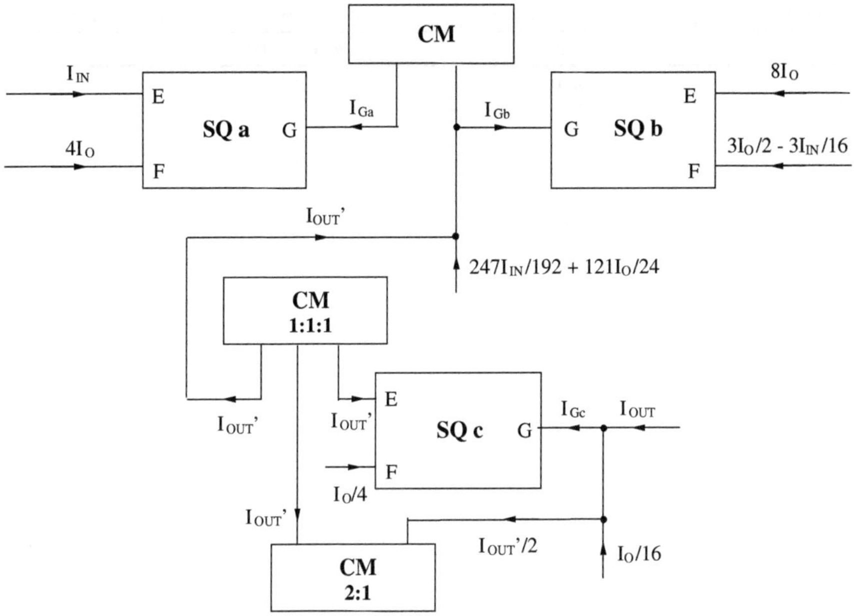

Fig. 1.49 Block diagram of the exponential function synthesizer circuit based on $g_{5b}(x)$ approximation function

A comparison between $g_{5c}(x)$ approximation function and $f(x) = \exp(x)$ function is presented in Table 1.15. The graphical representations of $g_{5c}(x)$ and $f(x) = \exp(x)$ functions are shown in Fig. 1.50, while the graphical representation of the approximation error, $\varepsilon(x)$, defined as the difference between $g_{5c}(x)$ and $f(x) = \exp(x)$ functions, is presented in Fig. 1.51.

The block diagram of the exponential function synthesizer circuit based on $g_{5c}(x)$ approximation function is presented in Fig. 1.52.

The expression of I_{Ga} and I_{Gb} currents are

$$I_{Ga} = 4I_O - \frac{I_{IN}}{2} + \frac{I_{IN}^2}{64I_O} \tag{1.131}$$

and

$$I_{Gb} = \frac{1}{4}\left(\frac{3I_O}{2} - \frac{3I_{IN}}{32}\right) - 4I_O + \frac{64I_O^2}{4\left(\frac{3I_O}{2} - \frac{3I_{IN}}{32}\right)}. \tag{1.132}$$

It results that I_{OUT}' current will have the following expression:

$$I_{OUT}' = I_{Gb} - I_{Ga} - \frac{343}{384}I_{IN} - \frac{49}{24}I_O, \tag{1.133}$$

which is equivalent to

Table 1.15 Comparison between $g_{5c}(x)$ approximation function and $f(x) = \exp(x)$ function

x	$f(x)$ (dB)	$g_{5c}(x)$ (dB)	ε (dB)
−2.4	−20.86	−22.81	1.95
−2.0	−17.38	−18.58	1.20
−1.6	−13.91	−14.60	0.69
−1.2	−10.43	−10.77	0.34
−0.8	−6.95	−7.09	0.14
−0.4	−3.48	−3.51	0.03
0.0	0.00	0.00	0.00
0.4	3.48	3.45	0.03
0.8	6.95	6.85	0.10
1.2	10.43	10.23	0.20
1.6	13.91	13.59	0.32
2.0	17.38	16.93	0.45
2.4	20.86	20.27	0.59
2.8	24.34	23.63	0.71
3.2	27.81	26.97	0.84
3.6	31.29	30.33	0.96
4.0	34.75	33.74	1.01
4.4	38.24	37.11	1.17
4.8	41.72	40.54	1.18
5.2	45.20	43.99	1.21
5.6	48.67	47.47	1.20
6.0	52.15	50.98	1.17
6.4	55.63	54.53	1.10
6.8	59.10	58.13	0.97
7.2	62.58	61.77	0.81
7.6	66.06	65.47	0.59
8.0	69.53	69.22	0.31
8.4	73.00	73.05	−0.05
8.8	76.48	76.95	−0.47
9.2	79.96	80.94	−0.98
9.6	83.44	85.04	−1.60
10.0	86.91	89.25	−2.34

$$I_{OUT}' = I_O\left[\frac{32}{3}\frac{1}{1-\frac{1}{16}\left(\frac{I_{IN}}{I_O}\right)} - \frac{1}{64}\left(\frac{I_{IN}}{I_O}\right)^2 - \frac{5}{12}\left(\frac{I_{IN}}{I_O}\right) - \frac{29}{3}\right]. \quad (1.134)$$

The I_{Gc} current can be expressed as follows:

$$I_{Gc} = \frac{I_O}{16} - \frac{I_{OUT}'}{2} + \frac{(I_{OUT}')^2}{I_O}, \quad (1.135)$$

while the expression of I_{OUT}'' current is

$$I_{OUT}'' = I_{Gc} - \frac{I_O}{16} + \frac{I_{OUT}'}{2}. \quad (1.136)$$

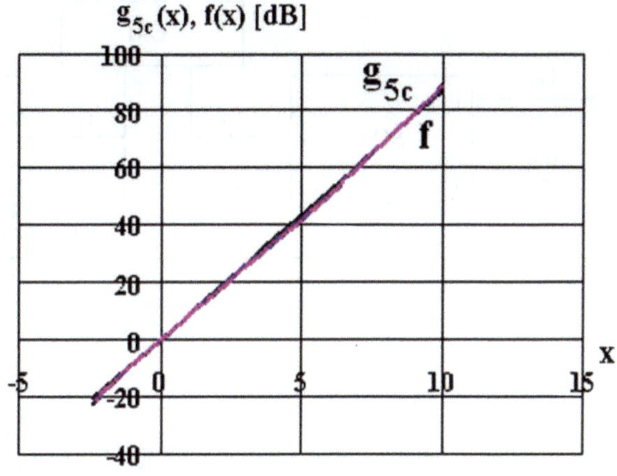

Fig. 1.50 Graphical representation of $g_{5c}(x)$ and $f(x) = \exp(x)$ functions

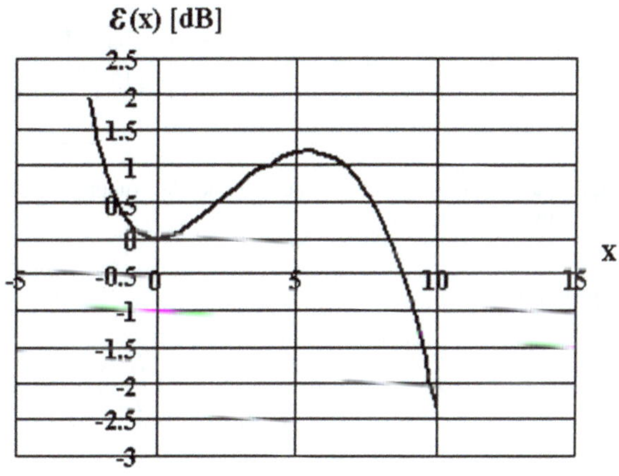

Fig. 1.51 Graphical representation of the approximation error, $\varepsilon(x)$

So

$$I_{OUT}'' = I_O \left[\frac{32}{3} \frac{1}{1 - \frac{1}{16}\left(\frac{I_{IN}}{I_O}\right)} - \frac{1}{64}\left(\frac{I_{IN}}{I_O}\right)^2 - \frac{5}{12}\left(\frac{I_{IN}}{I_O}\right) - \frac{29}{3} \right]^2. \tag{1.137}$$

The I_{Gd} current can be expressed as follows:

$$I_{Gd} = \frac{I_O}{16} - \frac{I_{OUT}''}{2} + \frac{(I_{OUT}'')^2}{I_O}, \tag{1.138}$$

resulting the following expression of I_{OUT} current:

$$I_{OUT} = I_{Gd} - \frac{I_O}{16} + \frac{I_{OUT}''}{2}, \tag{1.139}$$

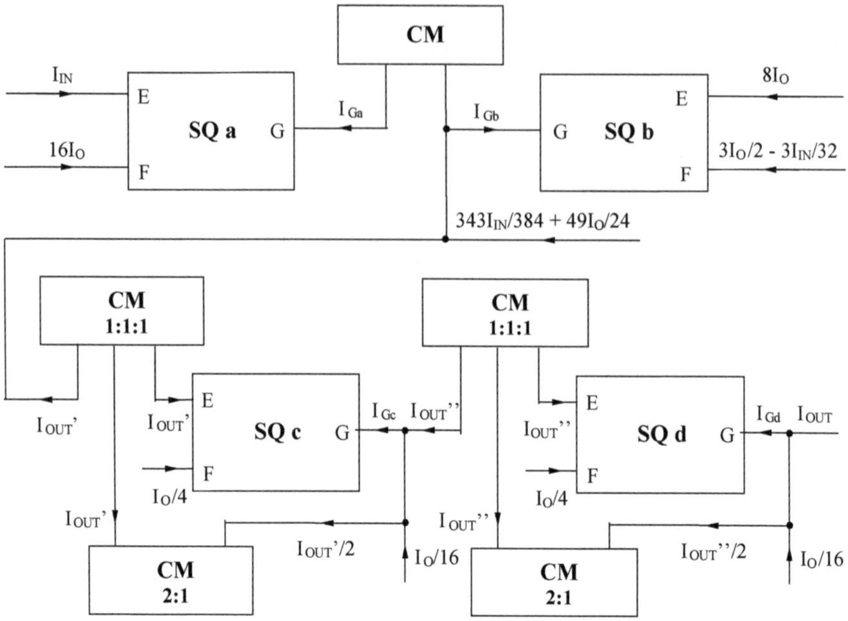

Fig. 1.52 Block diagram of the exponential function synthesizer circuit based on $g_{5c}(x)$ approximation function

equivalent to:

$$I_{\rm OUT} = I_O \left[\frac{32}{3} \frac{1}{1 - \frac{1}{16}\left(\frac{I_{\rm IN}}{I_O}\right)} - \frac{1}{64}\left(\frac{I_{\rm IN}}{I_O}\right)^2 - \frac{5}{12}\left(\frac{I_{\rm IN}}{I_O}\right) - \frac{29}{3} \right]^4. \tag{1.140}$$

In conclusion, $I_{\rm OUT}$ current approximates the exponential function using $g_{5c}(x)$ approximation function:

$$I_{\rm OUT} = I_O g_{5c}\left(\frac{I_{\rm IN}}{I_O}\right) \cong I_O \exp\left(\frac{I_{\rm IN}}{I_O}\right). \tag{1.141}$$

1.5.2 Second Class of Wide Output Dynamic Range Exponential Function Synthesizers with Fourth-Order Approximation

1.5.2.1 Approximation Function Without Variable Changing

Another fourth-order approximation function that uses two fractional functions and requires only two current-mode squaring circuits for its CMOS implementation can be expressed as follows:

$$g_{6a}(x) = \frac{1 + \mathrm{ax}}{1 + \mathrm{bx}} + \frac{c}{1 + x} + \mathrm{dx} + e. \tag{1.142}$$

The particular form of $g_{6a}(x)$ function, developed for approximating the exponential function, is

$$g_{6a}(x) = \frac{16 + 21x}{16 - 5x} - \frac{7}{11}x, \tag{1.143}$$

because $c \to 0$ and $e = 0$. A comparison between $g_{6a}(x)$ approximation function and $f(x) = \exp(x)$ function is presented in Table 1.16. The graphical representations of $g_{6a}(x)$ and $f(x) = \exp(x)$ functions are shown in Fig. 1.53, while the graphical representation of the approximation error, $\varepsilon(x)$, defined as the difference between $g_{6a}(x)$ and $f(x) = \exp(x)$ functions, is presented in Fig. 1.54.

The block diagram of the exponential function synthesizer circuit using $g_{6a}(x)$ approximation function is presented in Fig. 1.55. The "MD" block has the implementation presented in Fig. 1.3.

Table 1.16 Comparison between $g_{6a}(x)$ approximation function and $f(x) = \exp(x)$ function

x	$f(x)$ (dB)	$g_{6a}(x)$ (dB)	ε (dB)
−0.8	−8.06	−6.57	−0.38
−0.4	−3.65	−3.39	−0.09
0.0	0.00	0.00	0.00
0.4	3.40	3.45	0.03
0.8	6.74	6.94	0.01
1.2	10.13	10.52	−0.09
1.6	13.63	14.29	−0.38
2.0	17.30	18.48	−1.10
2.4	21.26	23.56	−2.70

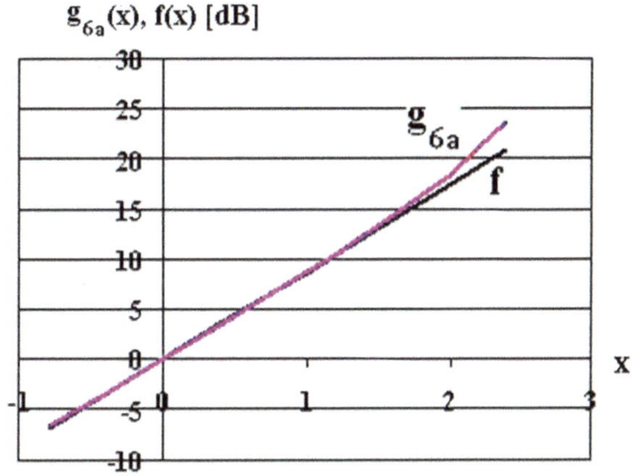

Fig. 1.53 Graphical representation of $g_{6a}(x)$ and $f(x) = \exp(x)$ functions

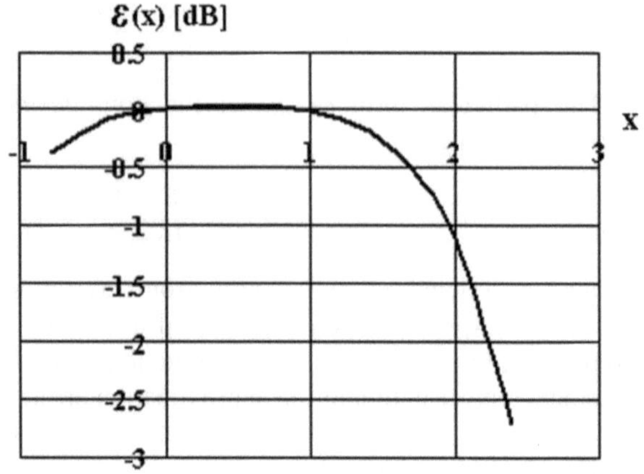

Fig. 1.54 Graphical representation of the approximation error, $\varepsilon(x)$

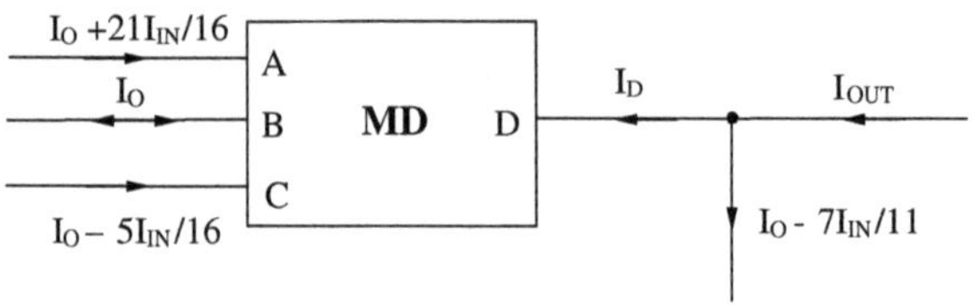

Fig. 1.55 Block diagram of the exponential function synthesizer circuit using $g_{6a}(x)$ approximation function

The expression of I_D current is

$$I_D = I_O \frac{16I_O + 21I_{\text{IN}}}{16I_O - 5I_{\text{IN}}} - I_O. \tag{1.144}$$

It results the following expression of I_{OUT} current:

$$I_{\text{OUT}} = I_D + I_O - \frac{7}{11} I_{\text{IN}}. \tag{1.145}$$

So

$$I_{\text{OUT}} = I_O \left[\frac{16 + 21\left(\frac{I_{\text{IN}}}{I_O}\right)}{16 - 5\left(\frac{I_{\text{IN}}}{I_O}\right)} - \frac{7}{11}\left(\frac{I_{\text{IN}}}{I_O}\right) \right]. \tag{1.146}$$

In conclusion, I_{OUT} current approximates the exponential function using $g_{6a}(x)$ approximation function:

$$I_{\text{OUT}} = I_O g_{6a}\left(\frac{I_{\text{IN}}}{I_O}\right) \cong I_O \exp\left(\frac{I_{\text{IN}}}{I_O}\right). \tag{1.147}$$

1.5.2.2 Approximation Function with First Variable Changing

A possible variable changing that can be used for increasing the output dynamic range of the exponential function synthesizer circuit is $x \to x/2$. The improved form of the fourth-order approximation function, based on this variable changing, can be expressed as follows:

$$g_{6b}(x) = \left(\frac{32+21x}{32-5x} - \frac{7}{22}x\right)^2. \tag{1.148}$$

The comparison between $g_{6b}(x)$ approximation function and $f(x) = \exp(x)$ function is presented in Table 1.17. The graphical representations of $g_{6b}(x)$ and $f(x) = \exp(x)$ functions are shown in Fig. 1.56, while the graphical representation of the approximation error, $\varepsilon(x)$, defined as the difference between $g_{6b}(x)$ and $f(x) = \exp(x)$ functions, is presented in Fig. 1.57.

Table 1.17 Comparison between $g_{6b}(x)$ approximation function and $f(x) = \exp(x)$ function

x	$f(x)$ (dB)	$g_{6b}(x)$ (dB)	ε (dB)
−2	−17.38	−15.99	−1.39
−1.6	−13.91	−13.15	−0.76
−1.2	−10.43	−10.05	−0.38
−0.8	−6.95	−6.78	−0.17
−0.4	−3.48	−3.42	−0.06
0.0	0.00	0.00	0.00
0.4	3.48	3.45	0.03
0.8	6.95	6.91	0.04
1.2	10.43	10.39	0.04
1.6	13.91	13.89	0.02
2.0	17.38	17.43	−0.05
2.4	20.86	21.03	0.17
2.8	24.34	24.74	−0.40
3.2	27.81	28.58	−0.77
3.6	31.29	32.62	−1.33
4.0	34.75	37.90	−3.15

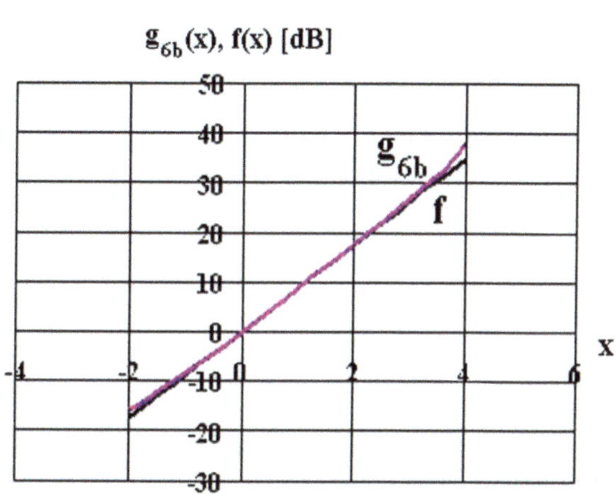

Fig. 1.56 Graphical representation of $g_{6b}(x)$ and $f(x) = \exp(x)$ functions

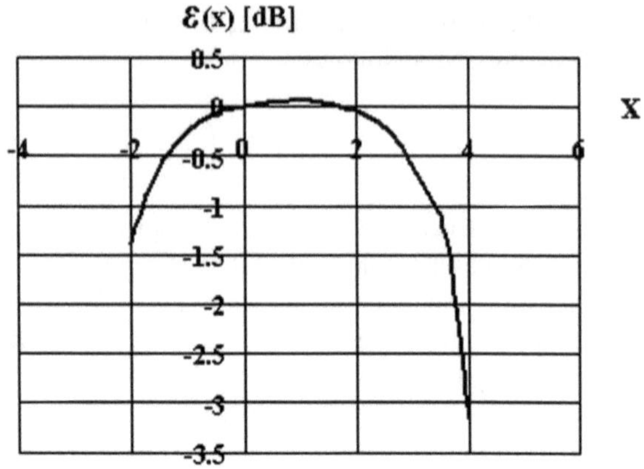

Fig. 1.57 Graphical representation of the approximation error, $\varepsilon(x)$

The exponential function synthesizer circuit based on $g_{6b}(x)$ approximation function has the block diagram presented in Fig. 1.58.

The expression of $I_D$current is

$$I_D = I_O \frac{I_O + \frac{21}{32} I_{\text{IN}}}{I_O - \frac{5}{32} I_{\text{IN}}} - I_O. \tag{1.149}$$

So, the expression of I_{OUT}' current will be

$$I_{\text{OUT}}' = I_D + I_O - \frac{7}{22} I_{\text{IN}}, \tag{1.150}$$

equivalent to

$$I_{\text{OUT}}' = I_O \left[\frac{32 + 21\left(\frac{I_{\text{IN}}}{I_O}\right)}{32 - 5\left(\frac{I_{\text{IN}}}{I_O}\right)} - \frac{7}{22}\left(\frac{I_{\text{IN}}}{I_O}\right) \right]. \tag{1.151}$$

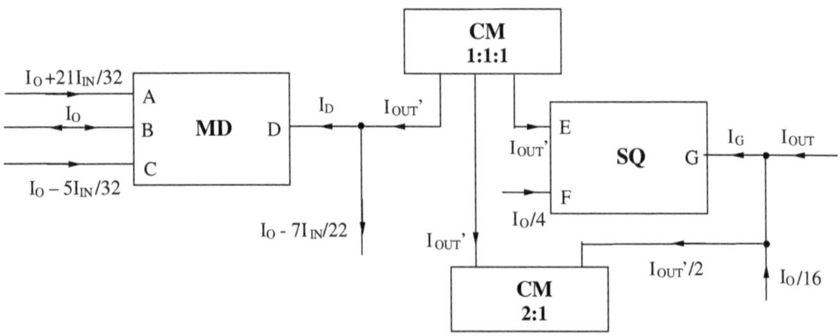

Fig. 1.58 Block diagram of the exponential function synthesizer circuit based on $g_{6b}(x)$ approximation function

The expression of I_G current can be written as

$$I_G = \frac{I_O}{16} - \frac{I_{OUT}'}{2} + \frac{(I_{OUT}')^2}{I_O}, \tag{1.152}$$

while the expression of I_{OUT} current is

$$I_{OUT} = I_G - \frac{I_O}{16} + \frac{I_{OUT}'}{2} = I_O \left[\frac{32 + 21\left(\frac{I_{IN}}{I_O}\right)}{32 - 5\left(\frac{I_{IN}}{I_O}\right)} - \frac{7}{22}\left(\frac{I_{IN}}{I_O}\right) \right]^2. \tag{1.153}$$

Thus, I_{OUT} current approximates the exponential function using $g_{6b}(x)$ approximation function:

$$I_{OUT} = I_O g_{6b}\left(\frac{I_{IN}}{I_O}\right) \cong I_O \exp\left(\frac{I_{IN}}{I_O}\right). \tag{1.154}$$

1.5.2.3 Approximation Function with Second Variable Changing

The second variable changing that can be implemented in CMOS technology with a small additional increasing of the circuit complexity (an additional current-mode squaring circuit) is $x \to x/4$, resulting the following improved approximation function:

$$g_{6c}(x) = \left(\frac{64 + 21x}{64 - 5x} - \frac{7}{44}x \right)^4. \tag{1.155}$$

A comparison between $g_{6c}(x)$ approximation function and $f(x) = \exp(x)$ function is presented in Table 1.18. The graphical representations of $g_{6c}(x)$ and $f(x) = \exp(x)$ functions are shown in Fig. 1.59, while the graphical representation of the approximation error, $\varepsilon(x)$, defined as the difference between $g_{6c}(x)$ and $f(x) = \exp(x)$ functions, is presented in Fig. 1.60.

The block diagram of the exponential function synthesizer circuit based on $g_{6c}(x)$ approximation function is presented in Fig. 1.61.

The expression of I_D current is

$$I_D = I_O \frac{64 + 21 I_{IN}}{64 - 5 I_{IN}} - I_O. \tag{1.156}$$

So, I_{OUT}' current will have the following expression:

$$I_{OUT}' = I_D + I_O - \frac{7}{44} I_{IN}. \tag{1.157}$$

equivalent to

$$I_{OUT}' = I_O \left[\frac{64 + 21\left(\frac{I_{IN}}{I_O}\right)}{64 - 5\left(\frac{I_{IN}}{I_O}\right)} - \frac{7}{44}\left(\frac{I_{IN}}{I_O}\right) \right]. \tag{1.158}$$

Table 1.18 Comparison between $g_{6c}(x)$ approximation function and $f(x) = \exp(x)$ function

x	$f(x)$ (dB)	$g_{6c}(x)$ (dB)	ε (dB)
−4.0	−34.75	−31.99	2.76
−3.6	−31.29	−29.22	2.07
−3.2	−27.81	−26.30	1.51
−2.8	−24.34	−23.25	1.09
−2.4	−20.86	−20.10	−0.76
−2.0	−17.38	−16.86	−0.52
−1.6	−13.91	−13.56	−0.35
−1.2	−10.43	−10.22	−0.21
−0.8	−6.95	−6.84	−0.11
−0.4	−3.48	−3.43	−0.05
0.0	0.00	0.00	0.00
0.4	3.48	3.44	0.04
0.8	6.95	6.89	0.06
1.2	10.43	10.35	0.08
1.6	13.91	13.82	0.09
2.0	17.38	17.29	0.09
2.4	20.86	20.77	0.09
2.8	24.34	24.27	0.07
3.2	27.81	27.77	0.04
3.6	31.29	31.30	−0.01
4.0	34.75	34.86	−0.11
4.4	38.24	38.44	−0.20
4.8	41.72	42.07	−0.35
5.2	45.20	45.74	−0.54
5.6	48.67	49.47	−0.80
6.0	52.15	53.27	−1.12
6.4	55.63	57.16	−1.53
6.8	59.10	61.14	−2.04
7.2	62.58	65.25	−2.67

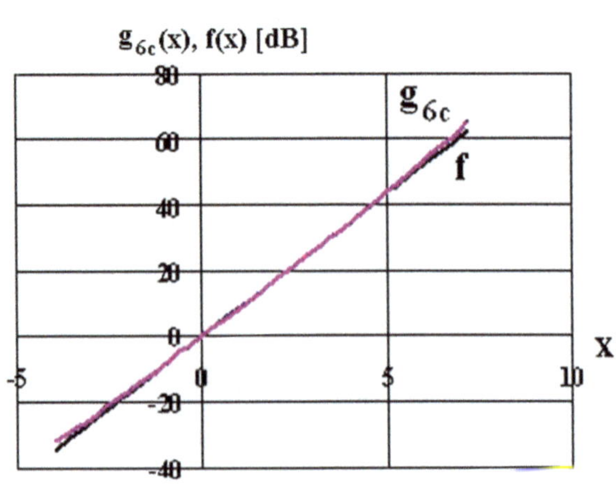

Fig. 1.59 Graphical representation of $g_{6c}(x)$ and $f(x) = \exp(x)$ functions

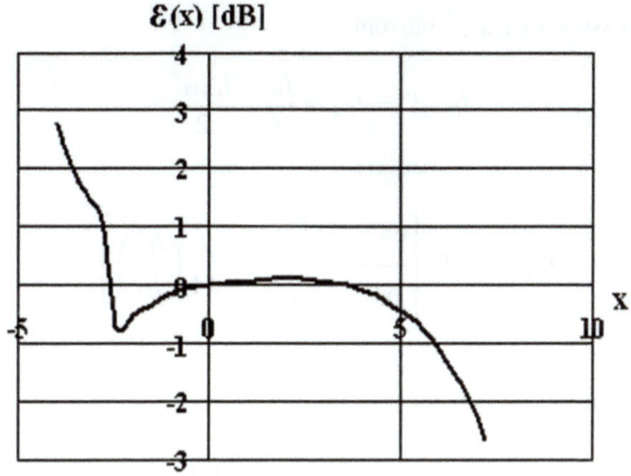

Fig. 1.60 Graphical representation of the approximation error, $\varepsilon(x)$

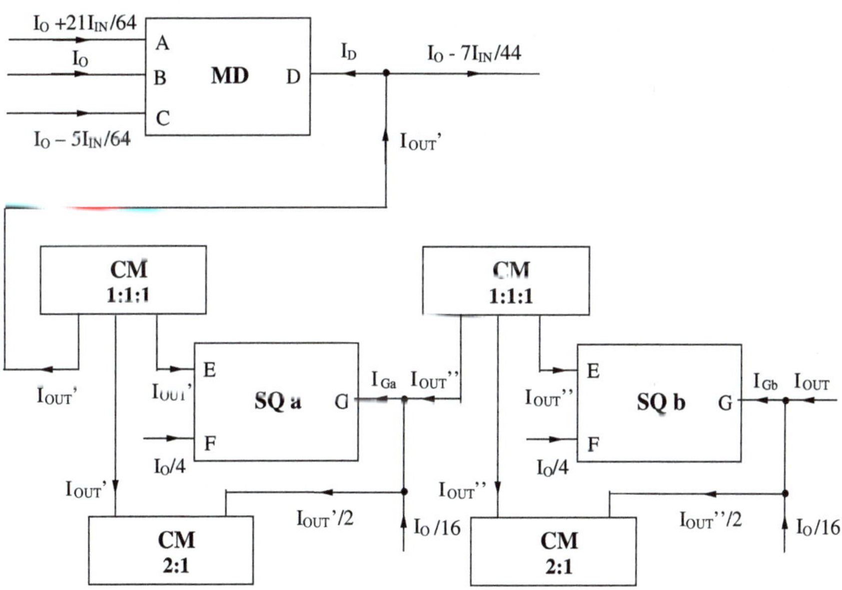

Fig. 1.61 Block diagram of the exponential function synthesizer circuit based on $g_{6c}(x)$ approximation function

The I_{Ga} current can be expressed as follows:

$$I_{Ga} = \frac{I_O}{16} - \frac{I_{OUT}'}{2} + \frac{(I_{OUT}')^2}{I_O}, \tag{1.159}$$

while the expression of I_{OUT}'' current is

$$I_{OUT}'' = I_{Ga} - \frac{I_O}{16} + \frac{I_{OUT}'}{2}. \tag{1.160}$$

It results:

$$I_{OUT}'' = I_O \left[\frac{64 + 21\left(\frac{I_{IN}}{I_O}\right)}{64 - 5\left(\frac{I_{IN}}{I_O}\right)} - \frac{7}{44}\left(\frac{I_{IN}}{I_O}\right) \right]^2. \tag{1.161}$$

The expression of I_{Gb} current will be:

$$I_{Gb} = \frac{I_O}{16} - \frac{I_{OUT}''}{2} + \frac{(I_{OUT}'')^2}{I_O}. \tag{1.162}$$

So, the I_{OUT} current can be expressed as follows:

$$I_{OUT} = I_{Gb} - \frac{I_O}{16} + \frac{I_{OUT}''}{2} = I_O \left[\frac{64 + 21\left(\frac{I_{IN}}{I_O}\right)}{64 - 5\left(\frac{I_{IN}}{I_O}\right)} - \frac{7}{44}\left(\frac{I_{IN}}{I_O}\right) \right]^4. \tag{1.163}$$

It results that I_{OUT} current approximates the exponential function using $g_{6c}(x)$ approximation function:

$$I_{OUT} = I_O g_{6c}\left(\frac{I_{IN}}{I_O}\right) \cong I_O \exp\left(\frac{I_{IN}}{I_O}\right). \tag{1.164}$$

Reference

1. Cosmin Popa, "High-Accuracy Function Synthesizer Circuit with Applications in Signal Processing", EURASIP Journal on Advanced in Signal Processing, 2012

Chapter 2
Wide Output Dynamic Range Gaussian Function Synthesizers

2.1 Introduction

The Gaussian function is intensively used in many domains of analog signal processing: neural networks and algorithms, wavelet transform and pattern recognition, neuro-fuzzy and classification applications, on-chip unsupervised learning, back-propagation neural networks. Similar requirements with exponential circuits can be considered also for the design of performance Gaussian function synthesizer structures: improved accuracy, good frequency response, low-power operation, reasonable complexity. Additionally, in order to increase the domain of applications for the developed Gaussian computational structures, the range of their input variable must be as large as possible. Convenient variable changing can be used for improving the operation of Gaussian function synthesizers. From this point of view, it is possible to reduce the order of approximation as a consequence of the accuracy increasing that could be obtained by using previously mentioned variable changing.

The complexity of CMOS implementations for Gaussian function synthesizer circuits can be strongly reduced by developing the approximation functions in such a way that they require relatively simple computational components. From this perspective, the most convenient choice is represented by the utilization of current-mode squaring or multiplier/divider circuits.

The chapter will analyze a multitude of possible realizations of Gaussian function synthesizer circuits, based on particular superior-order approximation functions: fourth-order, sixth-order, and eighth-order approximation functions. In order to improve the area of operation of the developed Gaussian circuits, convenient variable changing will be considered for each analyzed approximation function. Analytical and graphical analysis will be performed for determining the performances of these superior-order approximation functions.

The Gaussian function can be expressed as follows:

$$f(x) = A\exp\left(-\frac{x^2}{2\sigma^2}\right), \tag{2.1}$$

C. R. Popa, *Current-Mode Analog Nonlinear Function Synthesizer Structures*,
DOI: 10.1007/978-3-319-01035-9_2,

A and σ being constants that define the amplitude and the width of the Gaussian function, respectively. The expansion in the Taylor series of the Gaussian function has the following general expression:

$$f(x) = 1 - \frac{x^2}{2\sigma^2} + \frac{x^4}{8\sigma^4} - \frac{x^6}{48\sigma^6} + \cdots. \quad (2.2)$$

In order to increase the output dynamic range of the circuits those generate the Gaussian function, the reduction in the range of the input variable can be used, implementing a variable changing expressed by

$$x \rightarrow \frac{x}{\alpha}. \quad (2.3)$$

So, a Gaussian function synthesizer based on the previous variable changing will exploit the following mathematical relation:

$$\exp\left(-x^2\right) = \exp\left[-\left(\frac{x}{\alpha}\right)^2\right]^{\alpha^2}. \quad (2.4)$$

In order to maintain the complexity of the Gaussian computational circuit at a reasonable level, the usual value of α parameter is 2, resulting

$$\exp\left(-x^2\right) = \exp\left[-\left(\frac{x}{2}\right)^2\right]^4. \quad (2.5)$$

The method presented in this chapter for generating the Gaussian function is based on the utilization of superior-order approximation functions. Because the Gaussian function is an even-order function, the odd-order terms from its Taylor series expansions will be zero.

2.2 Fourth-Order Approximation of Gaussian Function

In order to generate the Gaussian function with a reasonable complexity of the function synthesizer circuit, a fourth-order approximation function can be used. It is necessary to make a trade-off between the accuracy of the approximation and the complexity of the Gaussian function synthesizer. From this point of view, the fourth-order approximation represents a convenient choice.

2.2.1 Approximation Function Without Variable Changing

2.2.1.1 Implementation of the Particular Gaussian Function

Based on the previous considerations, a possible fourth-order approximation function [1] could be generally expressed as follows:

$$g_{1a}(x) = \frac{b}{1+ax} + \frac{c}{1+x} + dx + e, \tag{2.6}$$

a, b, c, d, and e being constant coefficients having the values imposed by the condition that $g_{1a}(x)$ approximation function should match, in a fourth-order approximation, the Gaussian function.

The superior-order Taylor series of the approximation function has the following expression [1]:

$$\begin{aligned} g_{1a}(x) &= (b+c+e) + (d-c-ab)x + \left(c+a^2b\right)x^2 \\ &\quad - \left(c+a^3b\right)x^3 + \left(c+a^4b\right)x^4 - \left(c+a^5b\right)x^5 + \cdots \end{aligned} \tag{2.7}$$

Because $f(x)$ is an even-order function, all the odd-order terms from its series expansion are zero. The fourth-order identity between the previous functions is equivalent with the necessity of simultaneously fulfilling the following five mathematical relations [1]:

$$b + c + e = 1, \tag{2.8}$$

$$d - c - ab = 0, \tag{2.9}$$

$$c + a^2b = -\frac{1}{2\sigma^2}, \tag{2.10}$$

$$c + a^3b = 0, \tag{2.11}$$

$$c + a^4b = \frac{1}{8\sigma^4}. \tag{2.12}$$

Solving this system, it results the following values for $a \div e$ coefficients from the general expression of the proposed $g_{1a}(x)$ approximation function [1]:

$$a = \frac{1}{4\sigma^2}, \tag{2.13}$$

$$b = \frac{32\sigma^4}{1-4\sigma^2}, \tag{2.14}$$

$$c = \frac{1}{2\sigma^2(4\sigma^2-1)}, \tag{2.15}$$

$$d = \frac{16\sigma^4 - 1}{2\sigma^2(1 - 4\sigma^2)} \tag{2.16}$$

and

$$e = \frac{64\sigma^6 - 1}{2\sigma^2(4\sigma^2 - 1)} + 1. \tag{2.17}$$

As a result, the $g_{1a}(x)$ function can be expressed replacing (2.13)–(2.17) in (2.6).

The expression of $g_{1a}(x)$ function that fourth-order approximates the particular $f(x) = \exp(-x^2)$ Gaussian function can be obtained for $A = 1$ and $\sigma = 1/\sqrt{2}$, resulting [1]:

$$g_{1a}(x) = -\frac{8}{1 + \frac{x}{2}} + \frac{1}{1 + x} - 3x + 8. \tag{2.18}$$

As the fifth-order term of the Taylor series expansion is zero, the approximation error is mainly given by the sixth-order term from the same expansion [1]:

$$\varepsilon_{f(x)}^{g_{1a}(x)} \cong \frac{1}{48\sigma^6 \exp\left(-\frac{x^2}{2\sigma^2}\right)} = \frac{1}{6\exp(-x^2)}. \tag{2.19}$$

A comparison between $f(x) - \exp(-x^2)$ and $g_{1a}(x)$ approximation functions is shown in Table 2.1.

The graphical representations of $f(x) = \exp(-x^2)$ and $g_{1a}(x)$ functions are shown in Fig. 2.1, while the graphical representation of the approximation error, $\varepsilon(x)$, defined as the difference between $g_{1a}(x)$ and $f(x) = \exp(-x^2)$ functions, is presented in Fig. 2.2.

The current-mode squaring circuit used for implementing the Gauusian function synthesizers is presented in Fig. 2.3 [2], and its symbolical representation is shown in Fig. 2.4.

The expression of the output current is

$$I_G = \frac{I_E^2}{16I_F}. \tag{2.20}$$

Table 2.1 Comparison between $f(x) - \exp(-x^2)$ and $g_{1a}(x)$ approximation functions

x	$f(x)$	$g_{1a}(x)$	ε
−0.4	0.852	0.867	0.015
−0.2	0.961	0.961	0.000
0.0	1.000	1.000	0.000
0.2	0.961	0.961	0.000
0.4	0.852	0.848	−0.004
0.6	0.698	0.671	−0.027
0.8	0.527	0.441	−0.086

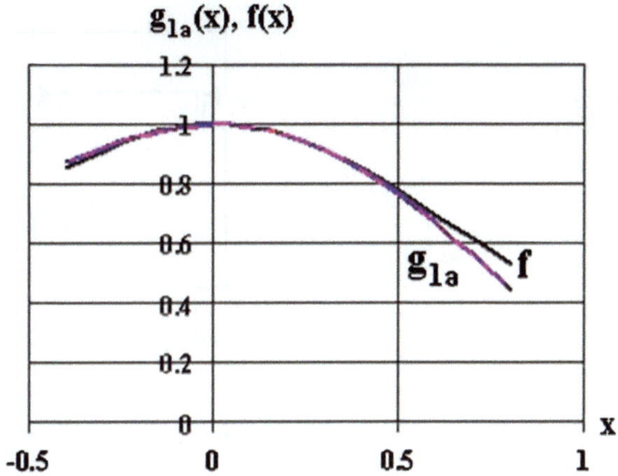

Fig. 2.1 Graphical representation of $f(x) = \exp(-x^2)$ and $g_{1a}(x)$ functions

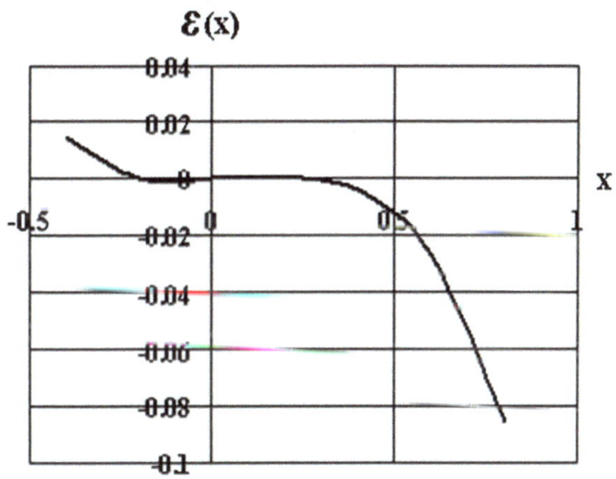

Fig. 2.2 Graphical representation of the approximation error, $\varepsilon(x)$

The block diagram of the particular Gaussian function synthesizer circuit that uses the $g_{1a}(x)$ approximation function and the implementation of the squaring circuit from Fig. 2.3 is presented in Fig. 2.5.

The expressions of I_{Ga} and I_{Gb} currents are:

$$I_{\text{Ga}} = \frac{8I_O^2}{I_O + \frac{I_{\text{IN}}}{2}} \tag{2.21}$$

and

$$I_{\text{Gb}} = \frac{I_O^2}{I_O + I_{\text{IN}}}, \tag{2.22}$$

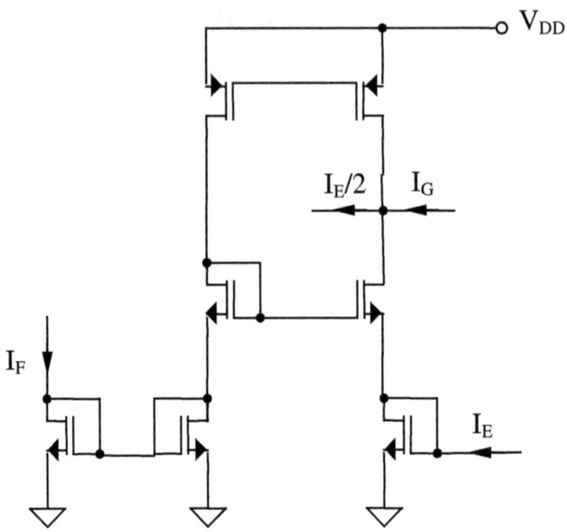

Fig. 2.3 Implementation of the current-mode squaring circuit [2]

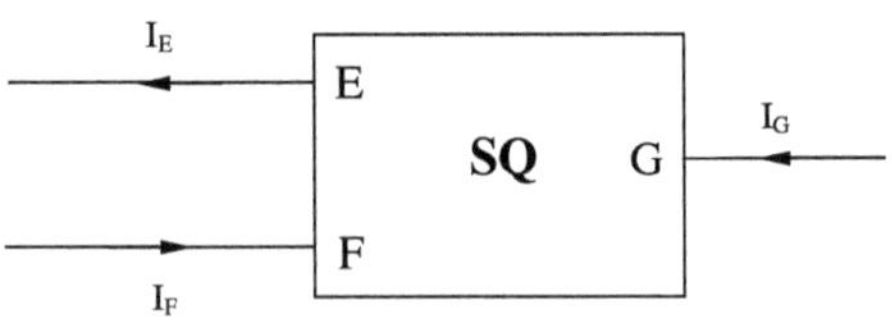

Fig. 2.4 Symbolical representation of the squaring circuit (*SQ*)

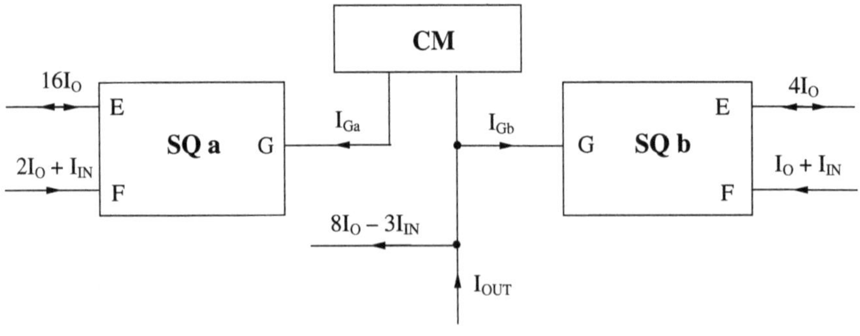

Fig. 2.5 Block diagram of the particular Gaussian function synthesizer circuit that uses the $g_{1a}(x)$ approximation function

resulting the following expression of I_{OUT} current:

$$I_{\text{OUT}} = I_{\text{Gb}} - I_{\text{Ga}} + 8I_O - 3I_{\text{IN}}, \tag{2.23}$$

equivalent with

$$I_{\text{OUT}} = I_O\left[-\frac{8}{1+\frac{1}{2}\left(\frac{I_{\text{IN}}}{I_O}\right)} + \frac{1}{1+\left(\frac{I_{\text{IN}}}{I_O}\right)} - 3\left(\frac{I_{\text{IN}}}{I_O}\right) + 8\right]. \tag{2.24}$$

So, I_{OUT} current approximates the particular Gaussian function using $g_{1a}(x)$ approximation function:

$$I_{\text{OUT}} = I_O g_{1a}\left(\frac{I_{\text{IN}}}{I_O}\right) \cong I_O \exp\left[-\left(\frac{I_{\text{IN}}}{I_O}\right)^2\right]. \tag{2.25}$$

2.2.1.2 Implementation of the General Gaussian Function

In order to implement the general Gaussian function $f(x) = A\exp\left(-x^2/2\sigma^2\right)$, the $x \to x/\sigma\sqrt{2}$ variable changing can be used, resulting the block diagram of the general Gaussian function synthesizer presented in Fig. 2.6.

The expressions of I_{Ga} and I_{Gb} currents are:

$$I_{\text{Ga}} = \frac{8I_O^2}{I_O + \frac{I_{\text{IN}}}{2\sigma\sqrt{2}}} \tag{2.26}$$

and

$$I_{\text{Gb}} = \frac{I_O^2}{I_O + \frac{I_{\text{IN}}}{\sigma\sqrt{2}}}, \tag{2.27}$$

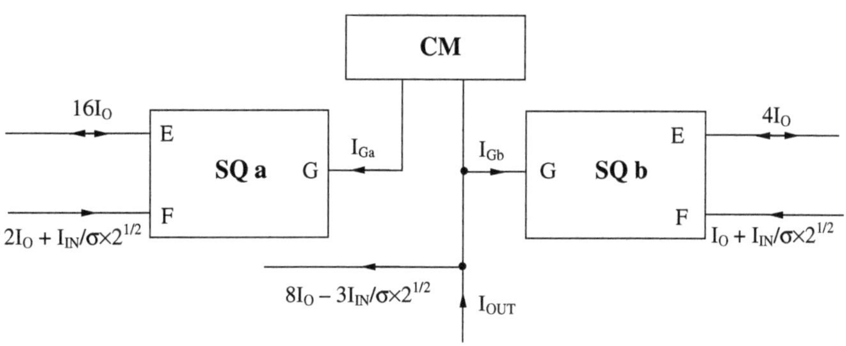

Fig. 2.6 Block diagram of the general Gaussian function synthesizer circuit that uses the $g_{1a}(x)$ approximation function

resulting the following expression of I_{OUT} current

$$I_{OUT} = I_{Gb} - I_{Ga} + 8I_O - \frac{3I_{IN}}{\sigma\sqrt{2}}, \tag{2.28}$$

equivalent with

$$I_{OUT} = I_O\left[-\frac{8}{1+\frac{1}{2\sigma\sqrt{2}}\left(\frac{I_{IN}}{I_O}\right)} + \frac{1}{1+\frac{1}{\sigma\sqrt{2}}\left(\frac{I_{IN}}{I_O}\right)} - \frac{3}{\sigma\sqrt{2}}\left(\frac{I_{IN}}{I_O}\right) + 8\right]. \tag{2.29}$$

So, I_{OUT} current approximates the general Gaussian function using $g_{1a}(x)$ approximation function:

$$I_{OUT} = I_O g_{1a}\left[\frac{1}{\sigma\sqrt{2}}\left(\frac{I_{IN}}{I_O}\right)\right] \cong I_O \exp\left[-\frac{1}{2\sigma^2}\left(\frac{I_{IN}}{I_O}\right)^2\right]. \tag{2.30}$$

2.2.2 *Approximation Function with Variable Changing*

2.2.2.1 Implementation of the Particular Gaussian Function

A possible method for increasing the output dynamic range of the Gaussian function synthesizer based on $g_{1a}(x)$ approximation function uses the $x \to x/2$ variable changing. The resulted approximation function, $g_{1b}(x)$, allows a relatively facile CMOS implementation, using two additional current-mode squaring circuits.

$$g_{1b}(x) = \left(-\frac{8}{1+\frac{x}{4}} + \frac{1}{1+\frac{x}{2}} - \frac{3x}{2} + 8\right)^4. \tag{2.31}$$

A comparison between $f(x) = \exp(-x^2)$ and $g_{1b}(x)$ approximation functions is shown in Table 2.2.

The graphical representations of $f(x) = \exp(-x^2)$ and $g_{1b}(x)$ functions are shown in Fig. 2.7, while the graphical representation of the approximation error, $\varepsilon(x)$, defined as the difference between $g_{1b}(x)$ and $f(x) = \exp(-x^2)$ functions, is presented in Fig. 2.8.

The block diagram of the particular Gaussian function synthesizer circuit that uses the $g_{1b}(x)$ approximation function is presented in Fig. 2.9.

The expressions of I_{Ga}, I_{Gb}, and I_{Gc} currents are

$$I_{Ga} = \frac{8I_O^2}{I_O + \frac{I_{IN}}{4}}, \tag{2.32}$$

Table 2.2 Comparison between $f(x) = \exp(-x^2)$ and $g_{1b}(x)$ approximation functions

x	$f(x)$	$g_{1b}(x)$	ε
−0.8	0.527	0.564	0.037
−0.6	0.698	0.706	0.008
−0.4	0.852	0.853	0.001
−0.2	0.961	0.961	0.000
0.0	1.000	1.000	0.000
0.2	0.961	0.961	0.000
0.4	0.852	0.851	−0.001
0.6	0.698	0.694	−0.004
0.8	0.527	0.516	−0.011
1.0	0.368	0.345	−0.023
1.2	0.237	0.203	−0.034

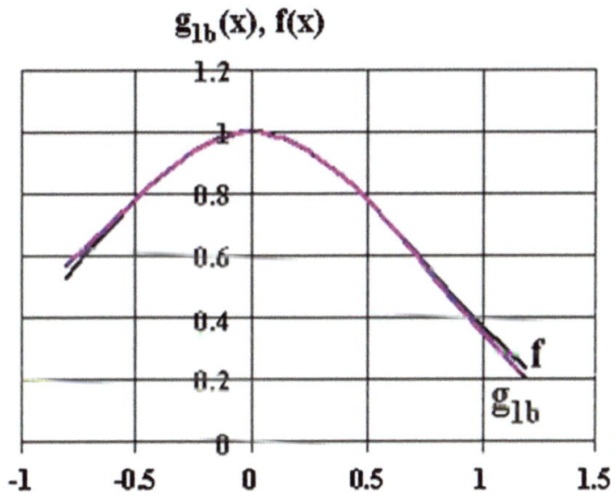

Fig. 2.7 Graphical representation of $f(x) = \exp(-x^2)$ and $g_{1b}(x)$ functions

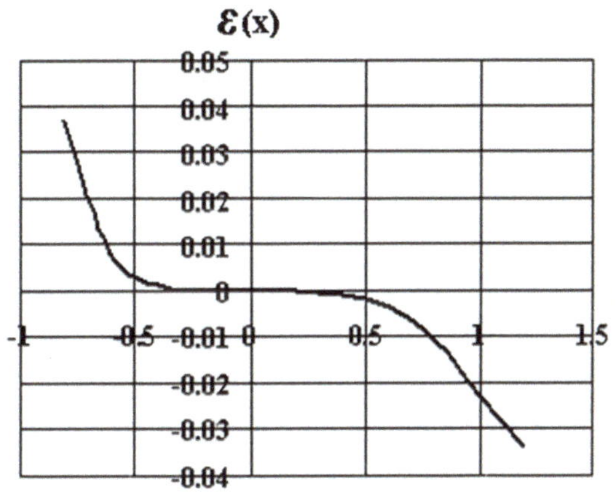

Fig. 2.8 Graphical representation of the approximation error, $\varepsilon(x)$

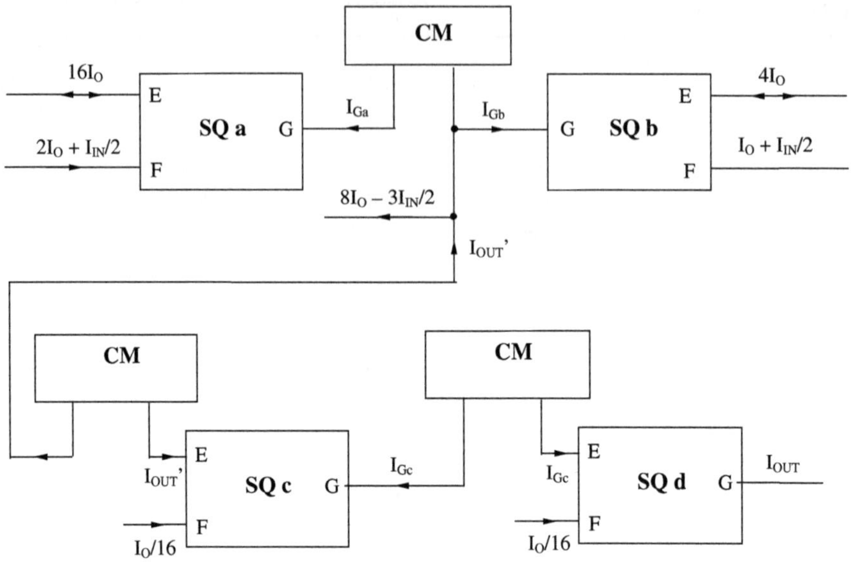

Fig. 2.9 Block diagram of the particular Gaussian function synthesizer circuit that uses the $g_{1b}(x)$ approximation function

$$I_{\mathrm{Gb}} = \frac{I_O^2}{I_O + \frac{I_{\mathrm{IN}}}{2}} \tag{2.33}$$

and

$$I_{\mathrm{Gc}} = \frac{(I_{\mathrm{OUT'}})^2}{I_O} = \frac{\left(I_{\mathrm{Gb}} - I_{\mathrm{Ga}} - \frac{3I_{\mathrm{IN}}}{2} + 8I_O\right)^2}{I_O}, \tag{2.34}$$

resulting

$$I_{\mathrm{Gc}} = I_O\left[-\frac{8}{1+\frac{1}{4}\left(\frac{I_{\mathrm{IN}}}{I_O}\right)} + \frac{1}{1+\frac{1}{2}\left(\frac{I_{\mathrm{IN}}}{I_O}\right)} - \frac{3}{2}\left(\frac{I_{\mathrm{IN}}}{I_O}\right) + 8\right]^2. \tag{2.35}$$

The I_{OUT} current can be expressed as follows:

$$I_{\mathrm{OUT}} = \frac{I_{\mathrm{Gc}}^2}{I_O} = I_O\left[-\frac{8}{1+\frac{1}{4}\left(\frac{I_{\mathrm{IN}}}{I_O}\right)} + \frac{1}{1+\frac{1}{2}\left(\frac{I_{\mathrm{IN}}}{I_O}\right)} - \frac{3}{2}\left(\frac{I_{\mathrm{IN}}}{I_O}\right) + 8\right]^4. \tag{2.36}$$

So, I_{OUT} current approximates the particular Gaussian function using $g_{1b}(x)$ approximation function:

$$I_{OUT} = I_O g_{1b}\left(\frac{I_{IN}}{I_O}\right) \cong I_O \exp\left[-\left(\frac{I_{IN}}{I_O}\right)^2\right]. \quad (2.37)$$

2.2.2.2 Implementation of the General Gaussian Function

Derived from the previous block diagram, presented for implementing the particular Gaussian function and using the previous variable changing, the block diagram of the general Gaussian function synthesizer circuit that uses the $g_{1b}(x)$ approximation function is presented in Fig. 2.10.

The expressions of I_{Ga}, I_{Gb}, and I_{Gc} currents are

$$I_{Ga} = \frac{8I_O^2}{I_O + \frac{I_{IN}}{4\sigma\sqrt{2}}}, \quad (2.38)$$

$$I_{Gb} = \frac{I_O^2}{I_O + \frac{I_{IN}}{2\sigma\sqrt{2}}} \quad (2.39)$$

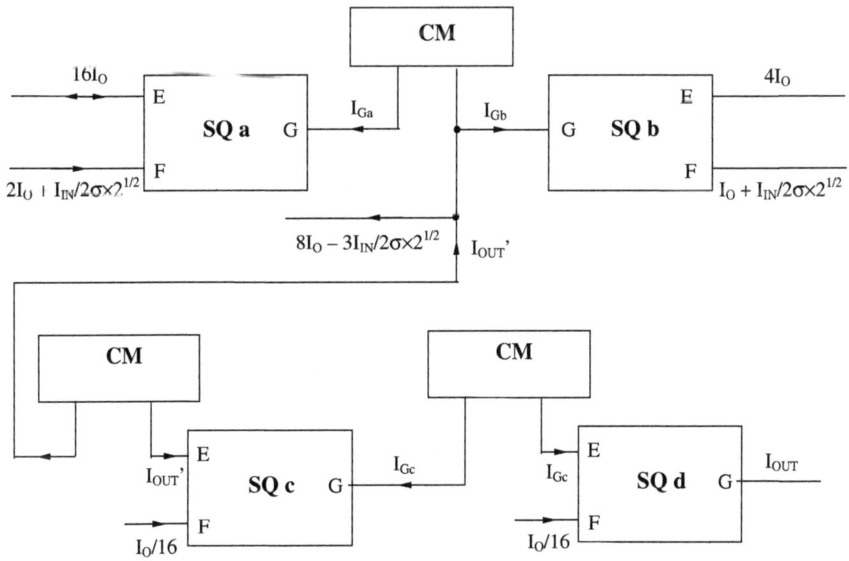

Fig. 2.10 Block diagram of the particular Gaussian function synthesizer circuit that uses the $g_{1b}(x)$ approximation function

and

$$I_{Gc} = \frac{(I_{OUT'})^2}{I_O} = \frac{\left(I_{Gb} - I_{Ga} - \frac{3I_{IN}}{2\sigma\sqrt{2}} + 8I_O\right)^2}{I_O}, \tag{2.40}$$

resulting

$$I_{Gc} = I_O\left[-\frac{8}{1+\frac{1}{4\sigma\sqrt{2}}\left(\frac{I_{IN}}{I_O}\right)} + \frac{1}{1+\frac{1}{2\sigma\sqrt{2}}\left(\frac{I_{IN}}{I_O}\right)} - \frac{3}{2\sigma\sqrt{2}}\left(\frac{I_{IN}}{I_O}\right) + 8\right]^2. \tag{2.41}$$

The expression of I_{OUT} current will be

$$I_{OUT} = \frac{I_{Gc}^2}{I_O} = I_O\left[-\frac{8}{1+\frac{1}{4\sigma\sqrt{2}}\left(\frac{I_{IN}}{I_O}\right)} + \frac{1}{1+\frac{1}{2\sigma\sqrt{2}}\left(\frac{I_{IN}}{I_O}\right)} - \frac{3}{2\sigma\sqrt{2}}\left(\frac{I_{IN}}{I_O}\right) + 8\right]^4. \tag{2.42}$$

Thus, I_{OUT} current approximates the particular Gaussian function using $g_{1b}(x)$ approximation function:

$$I_{OUT} = I_O g_{1b}\left[\frac{1}{\sigma\sqrt{2}}\left(\frac{I_{IN}}{I_O}\right)\right] \cong I_O \exp\left[-\frac{1}{2\sigma^2}\left(\frac{I_{IN}}{I_O}\right)^2\right]. \tag{2.43}$$

2.3 Sixth-Order Approximation of Gaussian Function Using Approximation Functions

In order to improve the accuracy of the Gaussian function synthesizer circuits, the order of approximation can be increased. This paragraph will present sixth-order Gaussian function synthesizer structures, developed in two cases: without or including variable changing for improving the circuit performances.

2.3.1 Approximation Function Without Variable Changing

2.3.1.1 Implementation of the Particular Gaussian Function

Based on the previous considerations, a possible general form of a sixth-order approximation function could be expressed as follows:

$$g_3(x) = \frac{b}{1+ax^2} + cx^2 + d, \tag{2.44}$$

Table 2.3 Comparison between $f(x) - \exp(-x^2)$ and $g_{2a}(x)$ approximation functions

x	$f(x)$	$g_{2a}(x)$	ε
−1.2	0.237	0.261	0.024
−1.0	0.368	0.375	0.007
−0.8	0.527	0.529	0.002
−0.6	0.698	0.698	0.000
−0.4	0.852	0.852	0.000
−0.2	0.961	0.961	0.000
0.0	1.000	1.000	0.000
0.2	0.961	0.961	0.000
0.4	0.852	0.852	0.000
0.6	0.698	0.698	0.000
0.8	0.527	0.529	0.002
1.0	0.368	0.375	0.007
1.2	0.237	0.261	0.024

a, b, c, and d being constant coefficients having the values imposed by the condition that $g_{2a}(x)$ approximation function should match, in a sixth-order approximation, the Gaussian function:

$$g_{2a}(x) = \frac{9}{2}\frac{1}{1+\frac{x^2}{3}} + \frac{x^2}{2} - \frac{7}{2}. \tag{2.45}$$

A comparison between $f(x) - \exp(-x^2)$ and $g_{2a}(x)$ approximation functions is shown in Table 2.3.

The graphical representations of $f(x) = \exp(-x^2)$ and $g_{2a}(x)$ functions are shown in Fig. 2.11, while the graphical representation of the approximation error, $\varepsilon(x)$, defined as the difference between $g_{2a}(x)$ and $f(x) = \exp(-x^2)$ functions, is presented in Fig. 2.12.

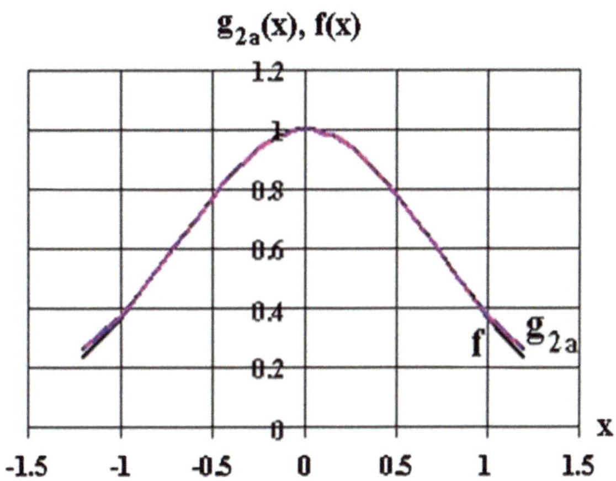

Fig. 2.11 Graphical representation of $f(x) = \exp(-x^2)$ and $g_{2a}(x)$ functions

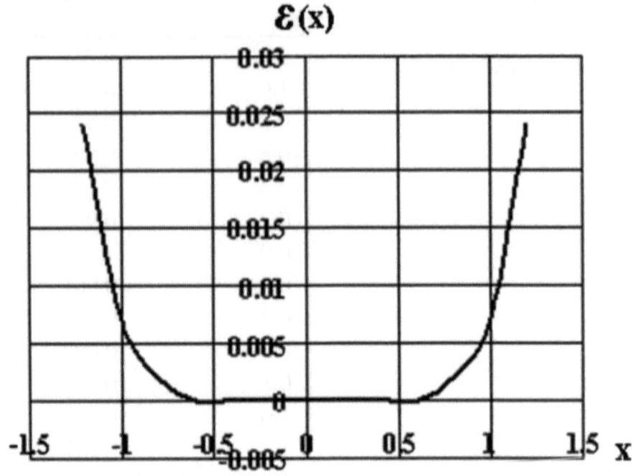

Fig. 2.12 Graphical representation of the approximation error, $\varepsilon(x)$

The block diagram of the particular Gaussian function synthesizer circuit that uses the $g_{2a}(x)$ approximation function is presented in Fig. 2.13.

The expressions of I_{Ga} and I_{Gb} currents are

$$I_{\mathrm{Ga}} = \frac{I_{\mathrm{IN}}^2}{3I_O}, \tag{2.46}$$

$$I_{\mathrm{Gb}} = \frac{72I_O^2}{16I_{\mathrm{OUT'}}} = \frac{9}{2}\frac{I_O^2}{I_{\mathrm{Ga}} + I_O} = \frac{9}{2}\frac{I_O^2}{\frac{I_{\mathrm{IN}}^2}{3I_O} + I_O}. \tag{2.47}$$

The I_{OUT} current can be expressed as follows:

$$I_{\mathrm{OUT}} = \frac{3}{2}I_{\mathrm{Ga}} + I_{\mathrm{Gb}} - \frac{7}{2}I_O, \tag{2.48}$$

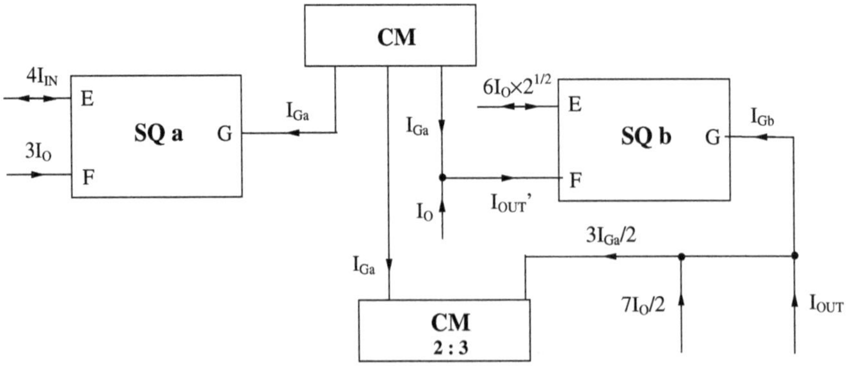

Fig. 2.13 Block diagram of the particular Gaussian function synthesizer circuit that uses the $g_{2a}(x)$ approximation function

resulting

$$I_{OUT} = I_O\left[\frac{9}{2}\frac{1}{1+\frac{1}{3}\left(\frac{I_{IN}}{I_O}\right)^2} + \frac{1}{2}\left(\frac{I_{IN}}{I_O}\right)^2 - \frac{7}{2}\right]. \tag{2.49}$$

So, I_{OUT} current approximates the particular Gaussian function using $g_{2a}(x)$ approximation function:

$$I_{OUT} = I_O g_{2a}\left(\frac{I_{IN}}{I_O}\right) \cong I_O \exp\left[-\left(\frac{I_{IN}}{I_O}\right)^2\right]. \tag{2.50}$$

2.3.1.2 Implementation of the General Gaussian Function

The block diagram of the general Gaussian function synthesizer circuit derived from the general form of $g_{2a}(x)$ approximation function is presented in Fig. 2.14.

The I_{Ga} and I_{Gb} currents can be expressed as follows:

$$I_{Ga} = \frac{I_{IN}^2}{6\sigma^2 I_O} \tag{2.51}$$

and

$$I_{Gb} = \frac{72 I_O^2}{16 I_{OUT'}} = \frac{9}{2}\frac{I_O^2}{I_{Ga}+I_O} = \frac{9}{2}\frac{I_O^2}{\frac{I_{IN}^2}{6\sigma^2 I_O}+I_O}. \tag{2.52}$$

The expression of I_{OUT} current will be

$$I_{OUT} = \frac{3}{2}I_{Ga} + I_{Gb} - \frac{7}{2}I_O, \tag{2.53}$$

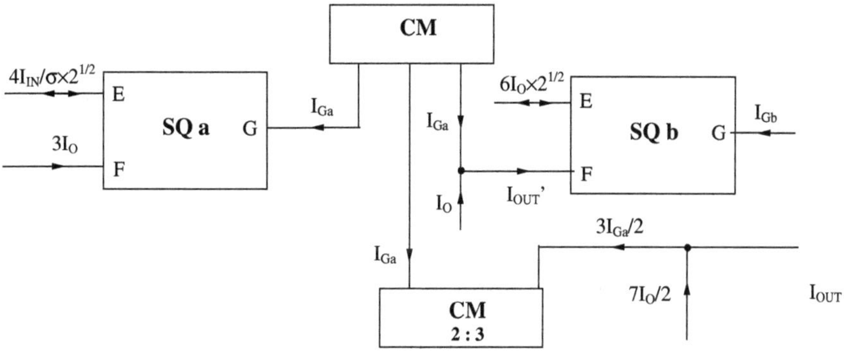

Fig. 2.14 Block diagram of the general Gaussian function synthesizer circuit that uses the $g_{2a}(x)$ approximation function

resulting

$$I_{OUT} = I_O \left[\frac{9}{2} \frac{1}{1 + \frac{1}{6\sigma^2} \left(\frac{I_{IN}}{I_O} \right)^2} + \frac{1}{4\sigma^2} \left(\frac{I_{IN}}{I_O} \right)^2 - \frac{7}{2} \right]. \tag{2.54}$$

It results that I_{OUT} current approximates the general Gaussian function using $g_{2a}(x)$ approximation function:

$$I_{OUT} = I_O g_{2a} \left[\frac{1}{\sigma\sqrt{2}} \left(\frac{I_{IN}}{I_O} \right) \right] \cong I_O \exp\left[-\frac{1}{2\sigma^2} \left(\frac{I_{IN}}{I_O} \right)^2 \right]. \tag{2.55}$$

2.3.2 *Approximation Function with Variable Changing*

2.3.2.1 Implementation of the Particular Gaussian Function

In order to improve the accuracy of the approximation and, in consequence, to increase the output dynamic range of the Gaussian function synthesizer based on $g_{2a}(x)$ approximation function, the $x \to x/2$ variable changing can be used. The resulted approximaton function, $g_{2b}(x)$, permits a relatively facile CMOS implementation, using only two additional current-mode squaring circuits.

$$g_{2b}(x) = \left(\frac{9}{2} \frac{1}{1 + \frac{x^2}{12}} + \frac{x^2}{8} - \frac{7}{2} \right)^4. \tag{2.56}$$

A comparison between $f(x) = \exp(-x^2)$ and $g_{2b}(x)$ approximation functions is shown in Table 2.4.

The graphical representations of $f(x) = \exp(-x^2)$ and $g_{2b}(x)$ functions are shown in Fig. 2.15. The graphical representation of the approximation error, $\varepsilon(x)$, defined as the difference between $g_{2b}(x)$ and $f(x) = \exp(-x^2)$ functions, is presented in Fig. 2.16.

The block diagram of the particular Gaussian function synthesizer circuit that uses the $g_{2b}(x)$ approximation function is presented in Fig. 2.17.

The expressions of I_{Ga} and I_{Gb} currents are

$$I_{Ga} = \frac{I_{IN}^2}{12 I_O} \tag{2.57}$$

and

$$I_{Gb} = \frac{72 I_O^2}{16 I_{OUT'}} = \frac{9}{2} \frac{I_O^2}{I_{Ga} + I_O} = \frac{9}{2} \frac{I_O^2}{\frac{I_{IN}^2}{12 I_O} + I_O}. \tag{2.58}$$

Table 2.4 Comparison between $f(x) = \exp(-x^2)$ and $g_{2b}(x)$ approximation functions

x	$f(x)$	$g_{2b}(x)$	ε
−2.4	0.003	0.005	0.002
−2.0	0.018	0.020	0.002
−1.6	0.077	0.078	0.001
−1.2	0.237	0.237	0.000
−0.8	0.527	0.527	0.000
−0.4	0.852	0.852	0.000
0.0	1.000	1.000	0.000
0.4	0.852	0.852	0.000
0.8	0.527	0.527	0.000
1.2	0.237	0.237	0.000
1.6	0.077	0.078	0.001
2.0	0.018	0.020	0.002
2.4	0.003	0.005	0.002

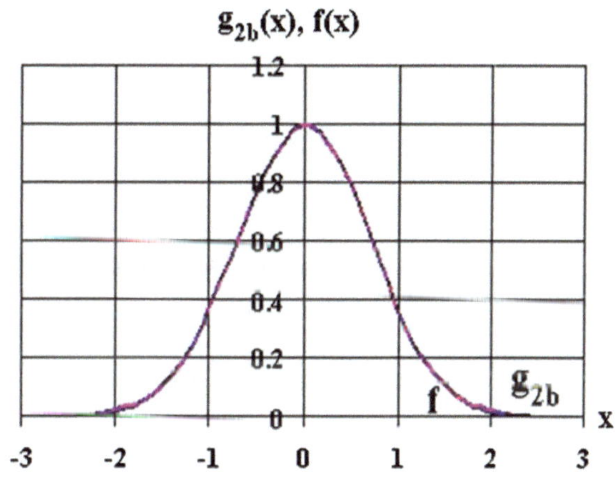

Fig. 2.15 Graphical representation of $f(x) = \exp(-x^2)$ and $g_{2b}(x)$ functions

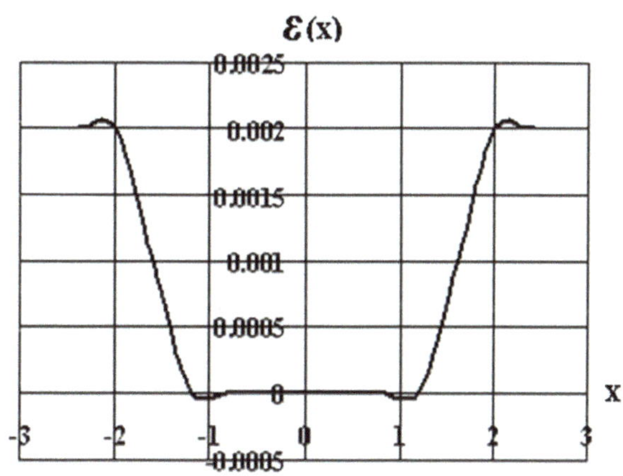

Fig. 2.16 Graphical representation of the approximation error, $\varepsilon(x)$

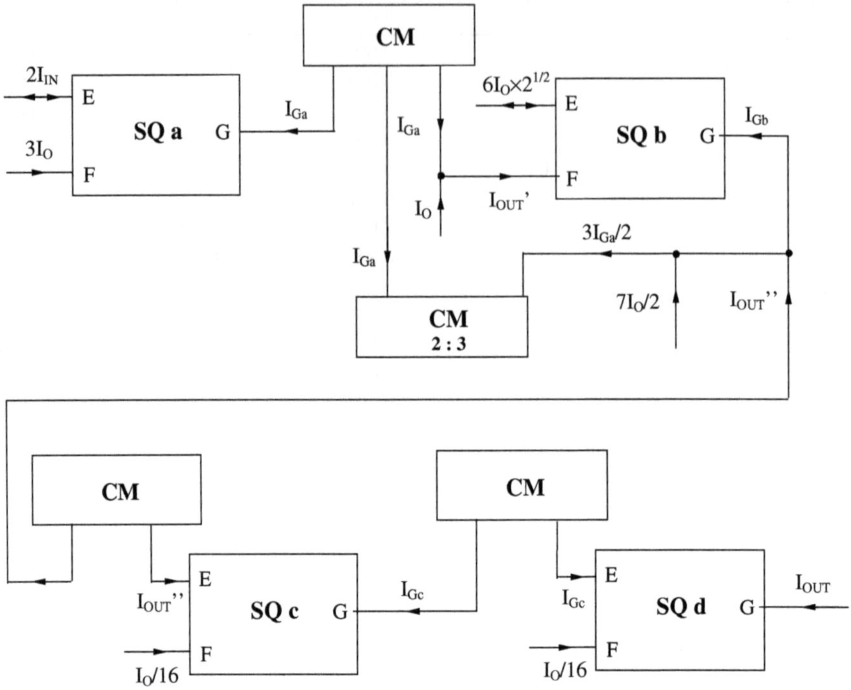

Fig. 2.17 Block diagram of the particular Gaussian function synthesizer circuit that uses the $g_{2b}(x)$ approximation function

The $I_{OUT''}$ current can be expressed as follows:

$$I_{OUT''} = \frac{3}{2} I_{Ga} + I_{Gb} - \frac{7}{2} I_O. \tag{2.59}$$

It results

$$I_{OUT''} = I_O \left[\frac{9}{2} \frac{1}{\left[1 + \frac{1}{12}\left(\frac{I_{IN}}{I_O}\right)\right]^2} + \frac{1}{8}\left(\frac{I_{IN}}{I_O}\right)^2 - \frac{7}{2} \right]. \tag{2.60}$$

The expression of I_{Gc} current is

$$I_{Gc} = \frac{(I_{OUT''})^2}{I_O} = I_O \left[\frac{9}{2} \frac{1}{\left[1 + \frac{1}{12}\left(\frac{I_{IN}}{I_O}\right)\right]^2} + \frac{1}{8}\left(\frac{I_{IN}}{I_O}\right)^2 - \frac{7}{2} \right]^2, \tag{2.61}$$

resulting the following expression of the output current:

$$I_{\mathrm{OUT}} = \frac{I_{\mathrm{Gc}}^2}{I_O} = I_O \left[\frac{9}{2} \frac{1}{1 + \frac{1}{12}\left(\frac{I_{\mathrm{IN}}}{I_O}\right)^2} + \frac{1}{8}\left(\frac{I_{\mathrm{IN}}}{I_O}\right)^2 - \frac{7}{2} \right]^4 . \tag{2.62}$$

So, I_{OUT} current approximates the particular Gaussian function using $g_{2b}(x)$ approximation function:

$$I_{\mathrm{OUT}} = I_O g_{2b}\left(\frac{I_{\mathrm{IN}}}{I_O}\right) \cong I_O \exp\left[-\left(\frac{I_{\mathrm{IN}}}{I_O}\right)^2 \right] . \tag{2.63}$$

2.3.2.2 Implementation of the General Gaussian Function

The block diagram of the general Gaussian function synthesizer circuit that uses the $g_{2b}(x)$ approximation function can be obtained from the previous block diagram using the $x \to x/\sigma\sqrt{2}$ variable changing (Fig. 2.18).

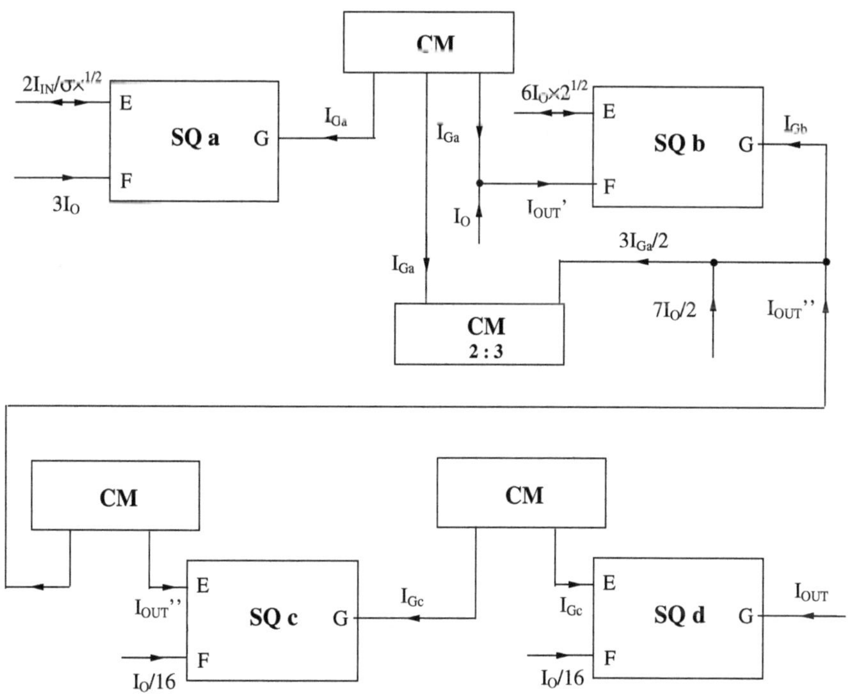

Fig. 2.18 Block diagram of the general Gaussian function synthesizer circuit that uses the $g_{2b}(x)$ approximation function

The I_{Ga} and I_{Gb} currents have the following expressions:

$$I_{Ga} = \frac{I_{IN}^2}{24\sigma^2 I_O} \tag{2.64}$$

and

$$I_{Gb} = \frac{72I_O^2}{16I_{OUT'}} = \frac{9}{2}\frac{I_O^2}{I_{Ga} + I_O} = \frac{9}{2}\frac{I_O^2}{\frac{I_{IN}^2}{24\sigma^2 I_O} + I_O}. \tag{2.65}$$

The $I_{OUT''}$ current can be expressed as follows:

$$I_{OUT''} = \frac{3}{2}I_{Ga} + I_{Gb} - \frac{7}{2}I_O, \tag{2.66}$$

resulting

$$I_{OUT''} = I_O\left[\frac{9}{2}\frac{1}{1 + \frac{1}{24\sigma^2}\left(\frac{I_{IN}}{I_O}\right)^2} + \frac{1}{16\sigma^2}\left(\frac{I_{IN}}{I_O}\right)^2 - \frac{7}{2}\right]. \tag{2.67}$$

The expression of I_{Gc} current is

$$I_{Gc} = \frac{(I_{OUT''})^2}{I_O} = I_O\left[\frac{9}{2}\frac{1}{1 + \frac{1}{24\sigma^2}\left(\frac{I_{IN}}{I_O}\right)^2} + \frac{1}{16\sigma^2}\left(\frac{I_{IN}}{I_O}\right)^2 - \frac{7}{2}\right]^2, \tag{2.68}$$

while the output current of the circuit can be expressed as follows:

$$I_{OUT} = \frac{I_{Gc}^2}{I_O} = I_O\left[\frac{9}{2}\frac{1}{1 + \frac{1}{24\sigma^2}\left(\frac{I_{IN}}{I_O}\right)^2} + \frac{1}{16\sigma^2}\left(\frac{I_{IN}}{I_O}\right)^2 - \frac{7}{2}\right]^4. \tag{2.69}$$

In conclusion, I_{OUT} current approximates the particular Gaussian function using $g_{2b}(x)$ approximation function:

$$I_{OUT} = I_O g_{2b}\left[\frac{1}{\sigma\sqrt{2}}\left(\frac{I_{IN}}{I_O}\right)\right] \cong I_O \exp\left[-\frac{1}{2\sigma^2}\left(\frac{I_{IN}}{I_O}\right)^2\right]. \tag{2.70}$$

2.4 Sixth-Order Approximation of Gaussian Function Using Limited Taylor Series

A possible implementation of a sixth-order approximation function for developing an improved accuracy Gaussian function synthesizer circuit uses the sixth-order Taylor series expansion of the Gaussian function. In order to additionally increase the approximation accuracy, a proper variable changing can be used.

2.4.1 Approximation Function Without Variable Changing

2.4.1.1 Implementation of the Particular Gaussian Function

The sixth-order approximation function that uses the limited Taylor expansion of the Gaussian function can be expressed as follows:

$$g_{3a}(x) = 1 - x^2 + \frac{x^4}{2} - \frac{x^6}{6}. \tag{2.71}$$

A comparison between $f(x) - \exp(-x^2)$ and $g_{3a}(x)$ approximation functions is shown in Table 2.5.

The graphical representations of $f(x) = \exp(-x^2)$ and $g_{3a}(x)$ functions are shown in Fig. 2.19, while the graphical representation of the approximation error, $\varepsilon(x)$, defined as the difference between $g_{3a}(x)$ and $f(x) = \exp(-x^2)$ functions, is presented in Fig. 2.20.

The block diagram of the particular Gaussian function synthesizer circuit that uses the $g_{3a}(x)$ approximation function is presented in Fig. 2.21.

The I_{Ga}, I_{Gb}, and I_{Gc} currents' expressions are

$$I_{Ga} = \frac{I_{IN}^2}{I_O}, \tag{2.72}$$

Table 2.5 Comparison between $f(x) - \exp(-x^2)$ and $g_{3a}(x)$ approximation functions

x	$f(x)$	$g_{3a}(x)$	ε
−1.0	0.368	0.333	−0.035
−0.8	0.527	0.521	−0.006
−0.6	0.698	0.697	−0.001
−0.4	0.852	0.852	0.000
−0.2	0.961	0.961	0.000
0.0	1.000	1.000	0.000
0.2	0.961	0.961	0.000
0.4	0.852	0.852	0.000
0.6	0.698	0.697	−0.001
0.8	0.527	0.521	−0.006
1.0	0.368	0.333	−0.035

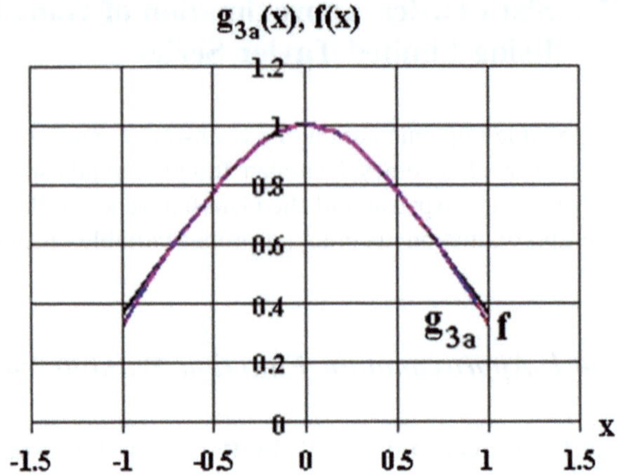

Fig. 2.19 Graphical representation of $f(x) = \exp(-x^2)$ and $g_{3a}(x)$ functions

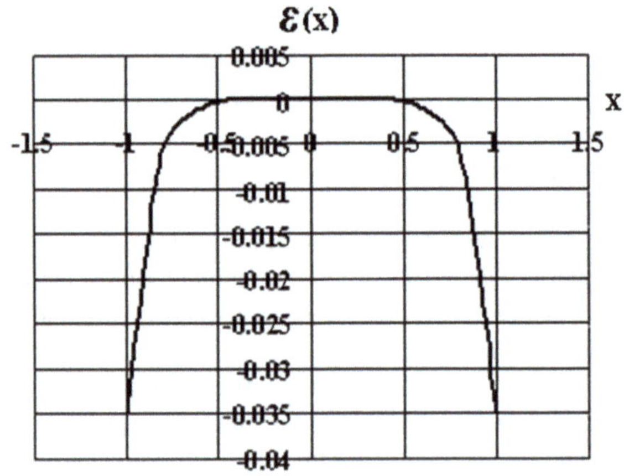

Fig. 2.20 Graphical representation of the approximation error, $\varepsilon(x)$

$$I_{Gb} = \frac{I_{G1}^2}{2I_O} = \frac{I_{IN}^4}{2I_O^3} \tag{2.73}$$

and

$$I_{Gc} = \frac{2}{3}\frac{I_{Gb}^2}{I_{Ga}} = \frac{I_{IN}^6}{6I_O^5}. \tag{2.74}$$

The I_{OUT} current can be expressed as follows:

$$I_{OUT} = I_{Gb} - I_{Ga} - I_{Gc} + I_O. \tag{2.75}$$

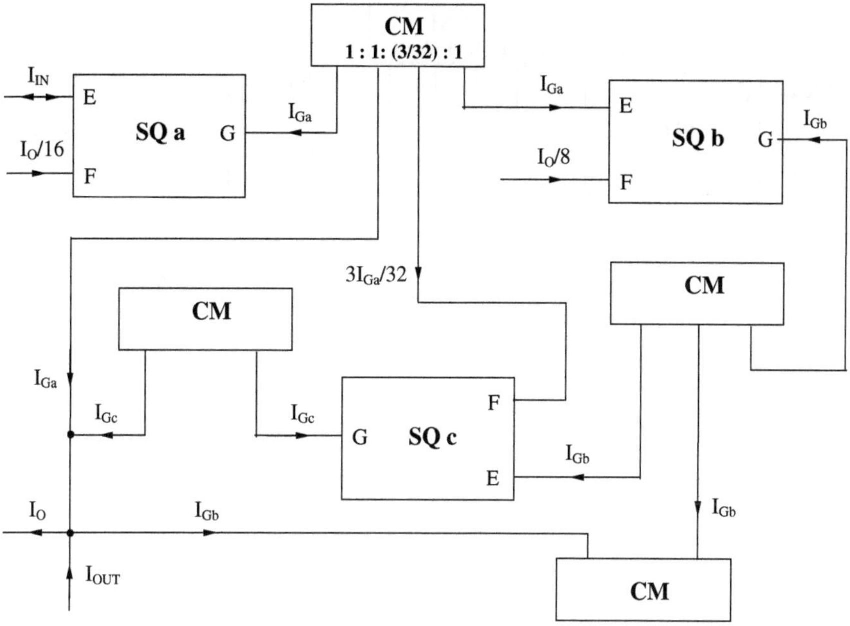

Fig. 2.21 Block diagram of the particular Gaussian function synthesizer circuit that uses the $g_{3a}(x)$ approximation function

It results

$$I_{OUT} = I_O\left[1 - \left(\frac{I_{IN}}{I_O}\right)^2 + \frac{1}{2}\left(\frac{I_{IN}}{I_O}\right)^4 - \frac{1}{6}\left(\frac{I_{IN}}{I_O}\right)^6\right]. \tag{2.76}$$

So, I_{OUT} current approximates the particular Gaussian function using $g_{3a}(x)$ approximation function:

$$I_{OUT} = I_O g_{3a}\left(\frac{I_{IN}}{I_O}\right) \cong I_O \exp\left[-\left(\frac{I_{IN}}{I_O}\right)^2\right]. \tag{2.77}$$

2.4.1.2 Implementation of the General Gaussian Function

The block diagram of the function synthesizer circuit that implements the general Gaussian function based on $g_{2b}(x)$ approximation function is presented in Fig. 2.22. The same variable changing was used, $x \to x/\sigma\sqrt{2}$.

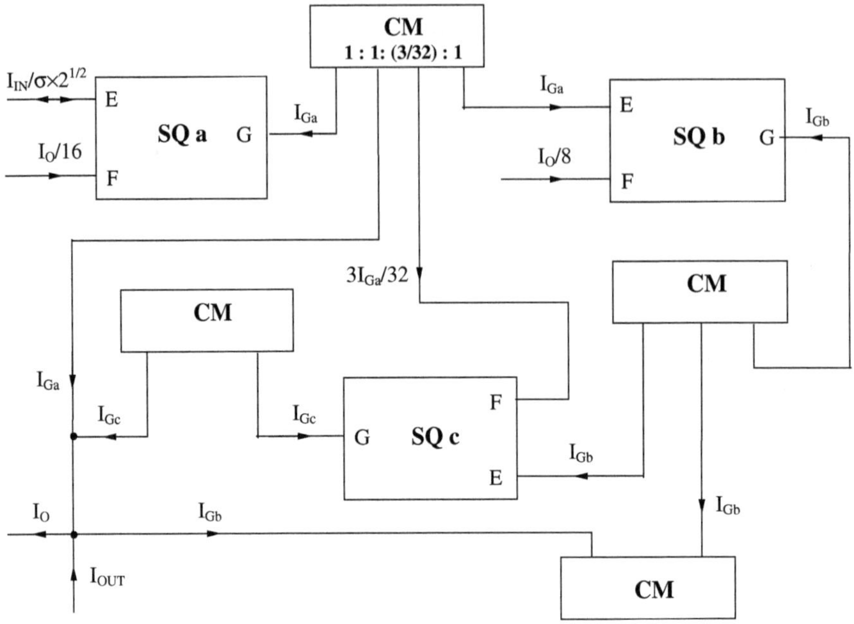

Fig. 2.22 Block diagram of the general Gaussian function synthesizer circuit that uses the $g_{3a}(x)$ approximation function

The expressions of I_{Ga}, I_{Gb}, and I_{Gc} currents are

$$I_{\text{Ga}} = \frac{I_{\text{IN}}^2}{2\sigma^2 I_O}, \tag{2.78}$$

$$I_{\text{Gb}} = \frac{I_{\text{G1}}^2}{2I_O} = \frac{I_{\text{IN}}^4}{8\sigma^4 I_O^3} \tag{2.79}$$

and

$$I_{\text{Gc}} = \frac{2}{3}\frac{I_{\text{Gb}}^2}{I_{\text{Ga}}} = \frac{I_{\text{IN}}^6}{48\sigma^6 I_O^5}. \tag{2.80}$$

The I_{OUT} current can be expressed as follows:

$$I_{\text{OUT}} = I_{\text{Gb}} - I_{\text{Ga}} - I_{\text{Gc}} + I_O, \tag{2.81}$$

resulting

$$I_{\text{OUT}} = I_O\left[1 - \frac{1}{2\sigma^2}\left(\frac{I_{\text{IN}}}{I_O}\right)^2 + \frac{1}{8\sigma^4}\left(\frac{I_{\text{IN}}}{I_O}\right)^4 - \frac{1}{48\sigma^6}\left(\frac{I_{\text{IN}}}{I_O}\right)^6\right]. \tag{2.82}$$

Thus, I_{OUT} current approximates the particular Gaussian function using $g_{3a}(x)$ approximation function:

$$I_{OUT} = I_O g_{3a}\left[\frac{1}{\sigma\sqrt{2}}\left(\frac{I_{IN}}{I_O}\right)\right] \cong I_O \exp\left[-\frac{1}{2\sigma^2}\left(\frac{I_{IN}}{I_O}\right)^2\right]. \tag{2.83}$$

2.4.2 *Approximation Function with Variable Changing*

2.4.2.1 Implementation of the Particular Gaussian Function

The increase in the Gaussian function synthesizer output dynamic range can be obtained, considering the $x \to x/2$ variable changing for developing the expression of $g_{3b}(x)$ approximation function. Using the $g_{3a}(x)$ function and the previously mentioned variable changing, the new approximation function can be expressed as follows:

$$g_{3b}(x) = \left(1 - \frac{x^2}{4} + \frac{x^4}{32} - \frac{x^6}{384}\right)^4. \tag{2.84}$$

A comparison between $f(x) = \exp(-x^2)$ and $g_{3b}(x)$ approximation functions is shown in Table 2.6.

Table 2.6 Comparison between $f(x) = \exp(-x^2)$ and $g_{3b}(x)$ approximation functions

x	$f(x)$	$g_{3b}(x)$	ε
−1.8	0.039	0.034	−0.005
−1.6	0.077	0.074	−0.003
−1.4	0.141	0.139	−0.002
−1.2	0.237	0.236	−0.001
−1.0	0.368	0.368	0.000
−0.8	0.527	0.527	0.000
−0.6	0.698	0.698	0.000
−0.4	0.852	0.852	0.000
−0.2	0.961	0.961	0.000
0.0	1.000	1.000	0.000
0.2	0.961	0.961	0.000
0.4	0.852	0.852	0.000
0.6	0.698	0.698	0.000
0.8	0.527	0.527	0.000
1.0	0.368	0.368	0.000
1.2	0.237	0.236	−0.001
1.4	0.141	0.139	−0.002
1.6	0.077	0.074	−0.003
1.8	0.039	0.034	−0.005

The graphical representations of $f(x) = \exp(-x^2)$ and $g_{3b}(x)$ functions are shown in Fig. 2.23, while the graphical representation of the approximation error, $\varepsilon(x)$, defined as the difference between $g_{3b}(x)$ and $f(x) = \exp(-x^2)$ functions, is presented in Fig. 2.24.

The block diagram of the particular Gaussian function synthesizer circuit that uses the $g_{3b}(x)$ approximation function is derived from the block diagram of the circuit using $g_{3a}(x)$ approximation function, having two additional current-mode squaring circuits. The output current of this Gaussian function synthesizer circuit will approximate the particular Gaussian function using $g_{3b}(x)$ approximation function:

$$I_{OUT} = I_O g_{3b}\left(\frac{I_{IN}}{I_O}\right) \cong I_O \exp\left[-\left(\frac{I_{IN}}{I_O}\right)^2\right]. \tag{2.85}$$

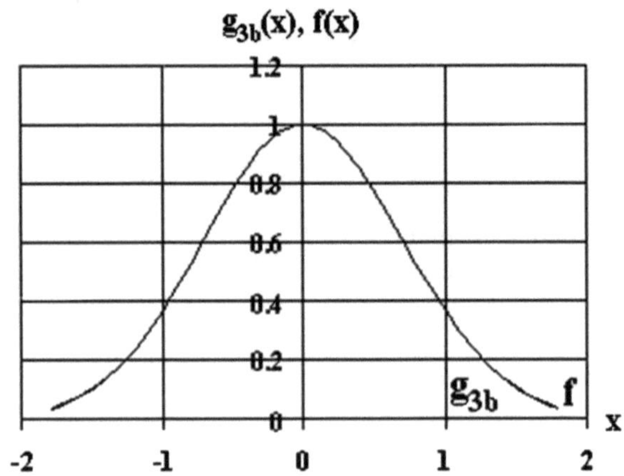

Fig. 2.23 Graphical representation of $f(x) = \exp(-x^2)$ and $g_{3b}(x)$ functions

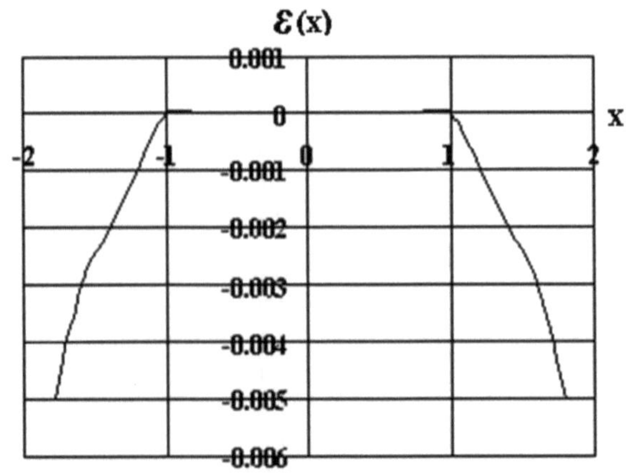

Fig. 2.24 Graphical representation of the approximation error, $\varepsilon(x)$

2.4.2.2 Implementation of the General Gaussian Function

Similar with the previous presented method, the $x \to x/\sigma\sqrt{2}$ variable changing allows to obtain an improved accuracy Gaussian function synthesizer circuit, based on the general form of the $g_{3b}(x)$ approximation function. The output current of the Gaussian function synthesizer structure, using the general form of $g_{3b}(x)$ approximation function, will accurately approximate the general Gaussian function:

$$I_{\text{OUT}} = I_O g_{3b}\left[\frac{1}{\sigma\sqrt{2}}\left(\frac{I_{\text{IN}}}{I_O}\right)\right] \cong I_O \exp\left[-\frac{1}{2\sigma^2}\left(\frac{I_{\text{IN}}}{I_O}\right)^2\right]. \tag{2.86}$$

2.5 Eighth-Order Approximation of Gaussian Function Using Approximation Functions

For applications that require extremely accurate generation of the Gaussian function, eighth-order approximation functions can be developed, taking into account similar restrictions comparing with previously designed computational structures.

2.5.1 Approximation Function Without Variable Changing

2.5.1.1 Implementation of the Particular Gaussian Function

The general form of a possible eighth-order approximation function could be expressed as follows:

$$g_{4a}(x) = \frac{b}{1 + ax^2} + cx^4 + dx^2 + e, \tag{2.87}$$

a, b, c, d, and e being constant coefficients having the values imposed by the condition that $g_{4a}(x)$ approximation function should match, in a sixth-order approximation, the Gaussian function:

$$g_{4a}(x) = \frac{32}{3}\frac{1}{1 + \frac{x^2}{4}} - \frac{x^4}{6} + \frac{5x^2}{3} - \frac{29}{3}. \tag{2.88}$$

A comparison between $f(x) - \exp(-x^2)$ and $g_{4a}(x)$ approximation functions is shown in Table 2.7.

Table 2.7 Comparison between $f(x) - \exp(-x^2)$ and $g_{4a}(x)$ approximation functions

x	$f(x)$	$g_{4a}(x)$	ε
−1.2	0.237	0.231	−0.006
−1.0	0.368	0.367	−0.001
−0.8	0.527	0.527	0.000
−0.6	0.698	0.698	0.000
−0.4	0.852	0.852	0.000
−0.2	0.961	0.961	0.000
0.0	1.000	1.000	0.000
0.2	0.961	0.961	0.000
0.4	0.852	0.852	0.000
0.6	0.698	0.698	0.000
0.8	0.527	0.527	0.000
1.0	0.368	0.367	−0.001
1.2	0.237	0.231	−0.006

The graphical representations of $f(x) = \exp(-x^2)$ and $g_{4a}(x)$ functions are shown in Fig. 2.25, while the graphical representation of the approximation error, $\varepsilon(x)$, defined as the difference between $g_{4a}(x)$ and $f(x) = \exp(-x^2)$ functions, is presented in Fig. 2.26.

The block diagram of the particular Gaussian function synthesizer circuit that uses the $g_{4a}(x)$ approximation function is presented in Fig. 2.27.

The expressions of I_{Ga}, I_{Gb}, and I_{Gc} currents are

$$I_{\text{Ga}} = \frac{I_{\text{IN}}^2}{4I_O}, \tag{2.89}$$

$$I_{\text{Gb}} = \frac{32I_O^2}{3I_{\text{OUT}'}} = \frac{32}{3}\frac{I_O^2}{I_{\text{Ga}} + I_O} = \frac{32}{3}\frac{I_O^2}{\frac{I_{\text{IN}}^2}{4I_O} + I_O} \tag{2.90}$$

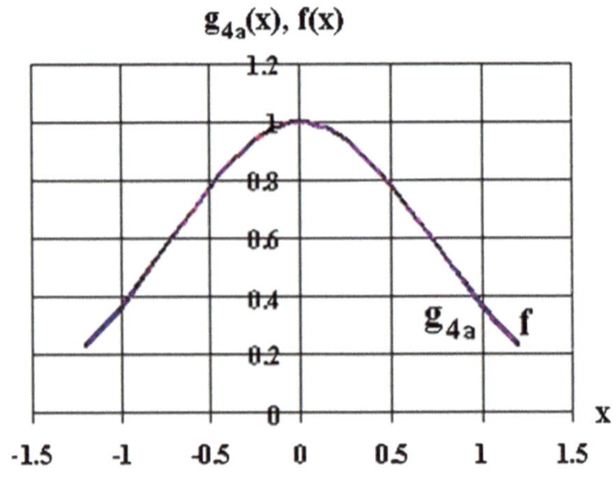

Fig. 2.25 Graphical representation of $f(x) = \exp(-x^2)$ and $g_{4a}(x)$ functions

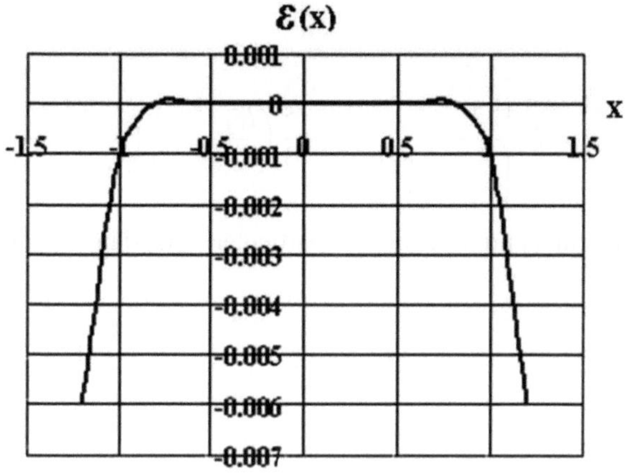

Fig. 2.26 Graphical representation of the approximation error, $\varepsilon(x)$

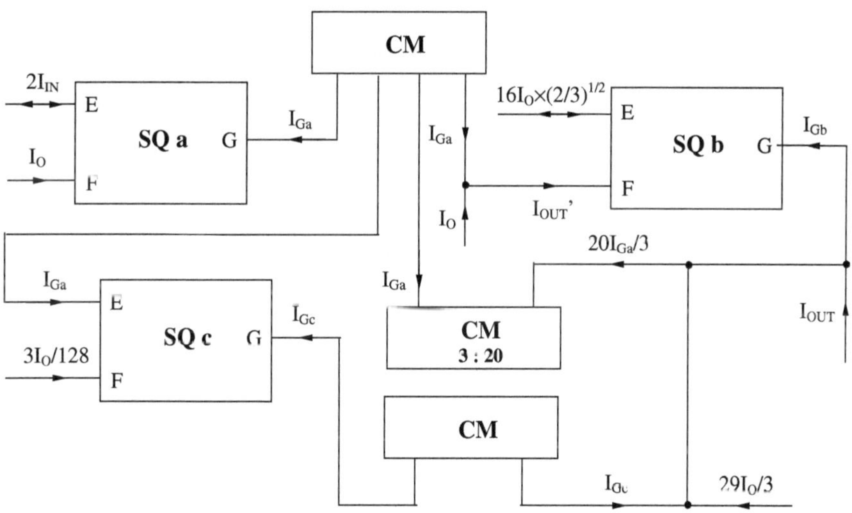

Fig. 2.27 Block diagram of the particular Gaussian function synthesizer circuit that uses the $g_{4a}(x)$ approximation function

and

$$I_{\mathrm{Gc}} = \frac{8}{3}\frac{I_{\mathrm{Ga}}^2}{I_O} = \frac{I_{\mathrm{IN}}^4}{6I_O^3}. \tag{2.91}$$

The I_{OUT} current expression will be

$$I_{\mathrm{OUT}} = I_{\mathrm{Gb}} + \frac{20}{3}I_{\mathrm{Ga}} - I_{\mathrm{Gc}} - \frac{29}{3}I_O, \tag{2.92}$$

resulting

$$I_{\text{OUT}} = I_O \left[\frac{32}{3} \frac{1}{1 + \frac{1}{4}\left(\frac{I_{\text{IN}}}{I_O}\right)^2} + \frac{5}{3}\left(\frac{I_{\text{IN}}}{I_O}\right)^2 - \frac{1}{6}\left(\frac{I_{\text{IN}}}{I_O}\right)^4 - \frac{29}{3} \right]. \tag{2.93}$$

So, I_{OUT} current approximates the particular Gaussian function using $g_{4a}(x)$ approximation function:

$$I_{\text{OUT}} = I_O g_{4a}\left(\frac{I_{\text{IN}}}{I_O}\right) \cong I_O \exp\left[-\left(\frac{I_{\text{IN}}}{I_O}\right)^2\right]. \tag{2.94}$$

2.5.1.2 Implementation of the General Gaussian Function

Using the previous function synthesizer circuit, developed for implementing the particular form of the $g_{4a}(x)$ approximation function, it is possible to design another Gaussian function synthesizer for implementing the general form of the same $g_{4a}(x)$ approximation function (Fig. 2.28).

The expressions of I_{Ga}, I_{Gb}, and I_{Gc} currents are

$$I_{\text{Ga}} = \frac{I_{\text{IN}}^2}{8\sigma^2 I_O}, \tag{2.95}$$

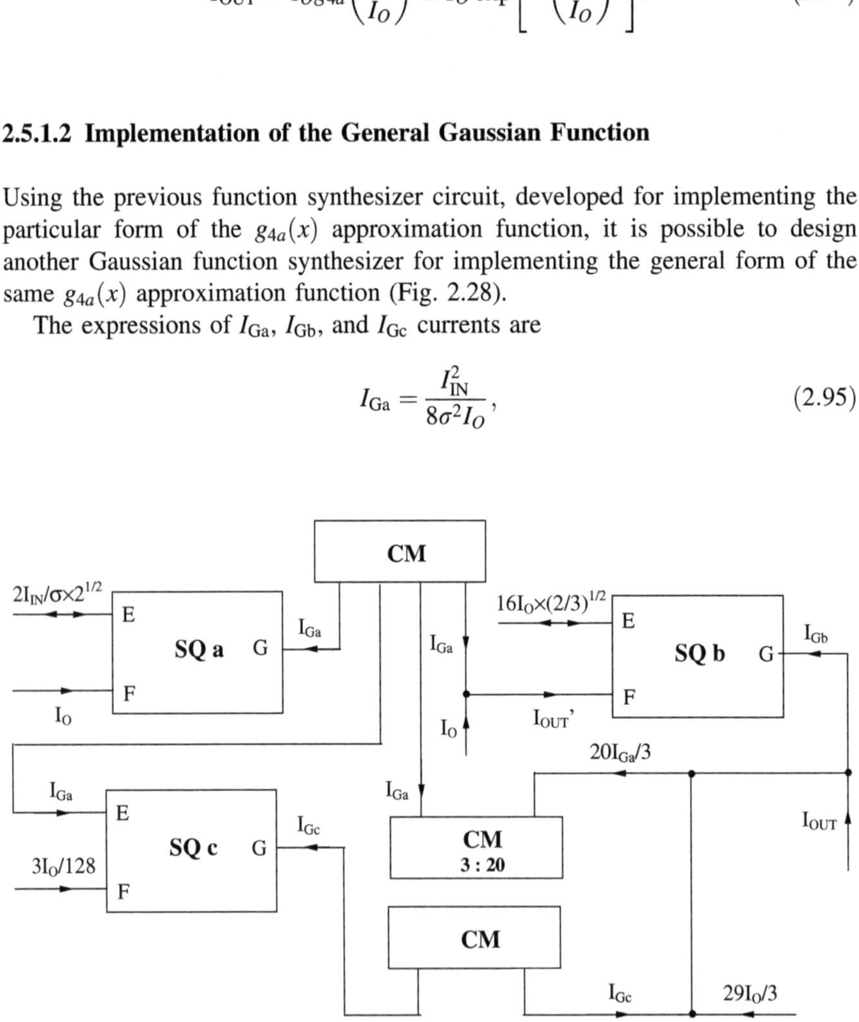

Fig. 2.28 Block diagram of the general Gaussian function synthesizer circuit that uses the $g_{4a}(x)$ approximation function

$$I_{\mathrm{Gb}} = \frac{32 I_O^2}{3 I_{\mathrm{OUT'}}} = \frac{32}{3} \frac{I_O^2}{I_{\mathrm{Ga}} + I_O} = \frac{32}{3} \frac{I_O^2}{\frac{I_{\mathrm{IN}}^2}{8\sigma^2 I_O} + I_O} \tag{2.96}$$

and

$$I_{\mathrm{Gc}} = \frac{8}{3} \frac{I_{\mathrm{Ga}}^2}{I_O} = \frac{I_{\mathrm{IN}}^4}{24\sigma^4 I_O^3}. \tag{2.97}$$

So, the I_{OUT} current will have the following expression:

$$I_{\mathrm{OUT}} = I_{\mathrm{Gb}} + \frac{20}{3} I_{\mathrm{Ga}} - I_{\mathrm{Gc}} - \frac{29}{3} I_O, \tag{2.98}$$

resulting

$$I_{\mathrm{OUT}} = I_O \left[\frac{32}{3} \frac{1}{1 + \frac{1}{8\sigma^2} \left(\frac{I_{\mathrm{IN}}}{I_O} \right)^2} + \frac{5}{6\sigma^2} \left(\frac{I_{\mathrm{IN}}}{I_O} \right)^2 - \frac{1}{24\sigma^4} \left(\frac{I_{\mathrm{IN}}}{I_O} \right)^4 - \frac{29}{3} \right]. \tag{2.99}$$

Thus, I_{OUT} current approximates the particular Gaussian function using $g_{4a}(x)$ approximation function:

$$I_{\mathrm{OUT}} = I_O g_{4a} \left[\frac{1}{\sigma\sqrt{2}} \left(\frac{I_{\mathrm{IN}}}{I_O} \right) \right] \cong I_O \exp \left[-\frac{1}{2\sigma^2} \left(\frac{I_{\mathrm{IN}}}{I_O} \right)^2 \right]. \tag{2.100}$$

2.5.2 *Approximation Function with Variable Changing*

2.5.2.1 Implementation of the Particular Gaussian Function

In order to obtain extremely accurate generation of the Gaussian function, additionally to the eighth-order of approximation, the classical $x \to x/2$ variable changing can be used. The increase in complexity for the resulted approximation function, $g_{4b}(x)$, is represented by only two additional current-mode squaring circuits:

$$g_{4b}(x) = \left(\frac{32}{3} \frac{1}{1 + \frac{x^2}{16}} - \frac{x^4}{96} + \frac{5x^2}{12} - \frac{29}{3} \right)^4. \tag{2.101}$$

A comparison between $f(x) = \exp(-x^2)$ and $g_{4b}(x)$ approximation functions is shown in Table 2.8.

The graphical representations of $f(x) = \exp(-x^2)$ and $g_{4b}(x)$ functions are shown in Fig. 2.29, while the graphical representation of the approximation error, $\varepsilon(x)$, defined as the difference between $g_{4b}(x)$ and $f(x) = \exp(-x^2)$ functions, is presented in Fig. 2.30.

Table 2.8 Comparison between $f(x) = \exp(-x^2)$ and $g_{4b}(x)$ approximation functions

x	$f(x)$	$g_{4b}(x)$	ε
−2.8	0.0004	0.0002	0.0002
−2.4	0.003	0.003	0.000
−2.0	0.018	0.018	0.000
−1.6	0.077	0.077	0.000
−1.2	0.237	0.237	0.000
−0.8	0.527	0.527	0.000
−0.4	0.852	0.852	0.000
0.0	1.000	1.000	0.000
0.4	0.852	0.852	0.000
0.8	0.527	0.527	0.000
1.2	0.237	0.237	0.000
1.6	0.077	0.077	0.000
2.0	0.018	0.018	0.000
2.4	0.003	0.003	0.000
2.8	0.0004	0.0002	−0.0002

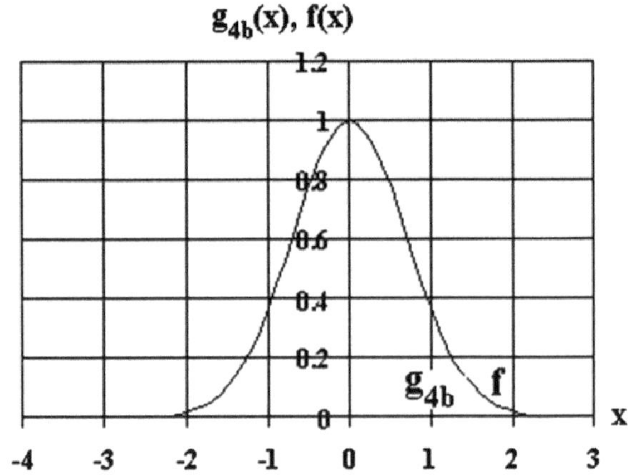

Fig. 2.29 Graphical representation of $f(x) = \exp(-x^2)$ and $g_{4b}(x)$ functions

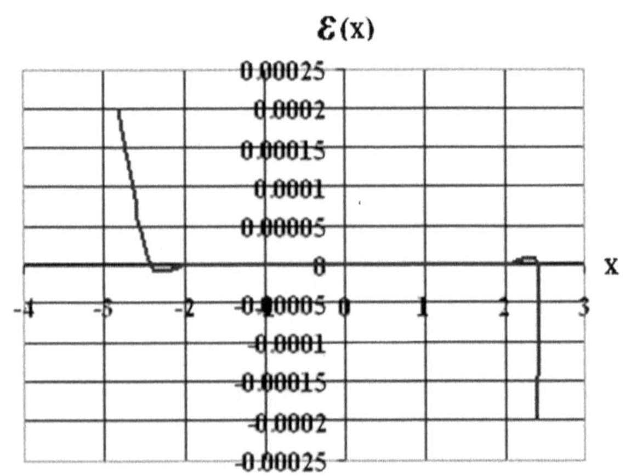

Fig. 2.30 Graphical representation of the approximation error, $\varepsilon(x)$

The block diagram of the particular Gaussian function synthesizer circuit that uses the $g_{4b}(x)$ approximation function is presented in Fig. 2.31.

The expressions of I_{Ga}, I_{Gb}, and I_{Gc} currents are

$$I_{Ga} = \frac{I_{IN}^2}{16I_O}, \tag{2.102}$$

$$I_{Gb} = \frac{32I_O^2}{3I_{OUT'}} = \frac{32}{3}\frac{I_O^2}{I_{G1} + I_O} = \frac{32}{3}\frac{I_O^2}{\frac{I_{IN}^2}{16I_O} + I_O} \tag{2.103}$$

and

$$I_{Gc} = \frac{8}{3}\frac{I_{G1}^2}{I_O} = \frac{I_{IN}^4}{96I_O^3}. \tag{2.104}$$

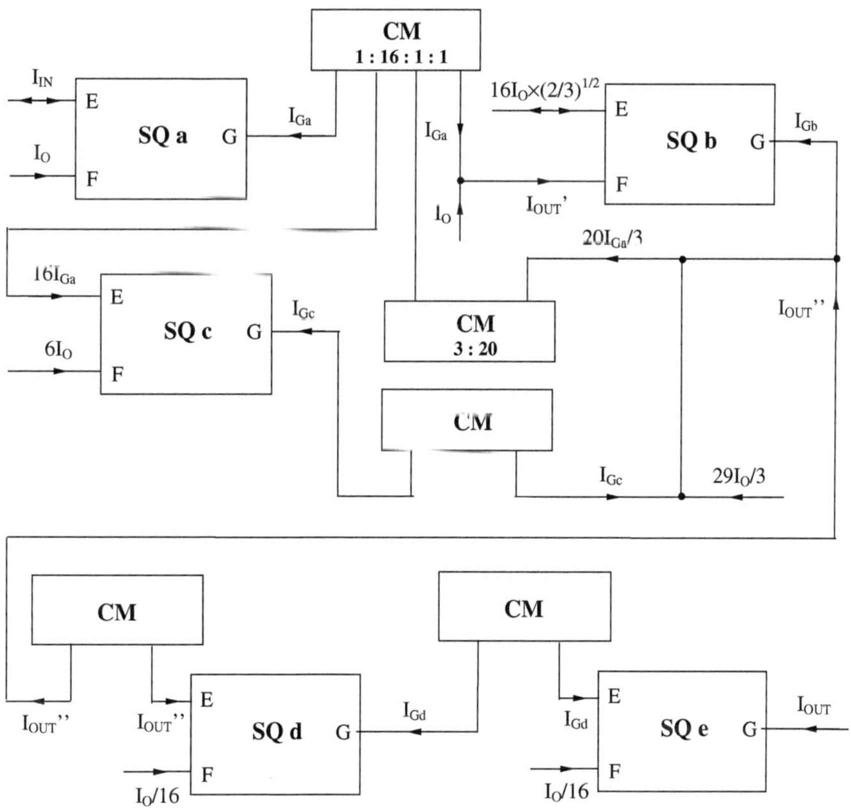

Fig. 2.31 Block diagram of the particular Gaussian function synthesizer circuit that uses the $g_{4b}(x)$ approximation function

The $I_{OUT''}$ current can be expressed as follows:

$$I_{\mathrm{OUT}''} = I_{\mathrm{Gb}} + \frac{20}{3} I_{\mathrm{Ga}} - I_{\mathrm{Gc}} - \frac{29}{3} I_O, \tag{2.105}$$

resulting

$$I_{\mathrm{OUT}''} = I_O \left[\frac{32}{3} \frac{1}{1 + \frac{1}{16}\left(\frac{I_{\mathrm{IN}}}{I_O}\right)^2} + \frac{5}{12}\left(\frac{I_{\mathrm{IN}}}{I_O}\right)^2 - \frac{1}{96}\left(\frac{I_{\mathrm{IN}}}{I_O}\right)^4 - \frac{29}{3} \right]. \tag{2.106}$$

The I_{Gd} and I_{OUT} currents have the following expressions:

$$I_{\mathrm{Gd}} = \frac{(I_{\mathrm{OUT}''})^2}{I_O} \tag{2.107}$$

and

$$I_{\mathrm{OUT}} = \frac{I_{\mathrm{Gd}}^2}{I_O} = \frac{(I_{\mathrm{OUT}''})^4}{I_O^3}. \tag{2.108}$$

Thus, the output current of the circuit can be expressed as follows:

$$I_{\mathrm{OUT}} = I_O \left[\frac{32}{3} \frac{1}{1 + \frac{1}{16}\left(\frac{I_{\mathrm{IN}}}{I_O}\right)^2} + \frac{5}{12}\left(\frac{I_{\mathrm{IN}}}{I_O}\right)^2 - \frac{1}{96}\left(\frac{I_{\mathrm{IN}}}{I_O}\right)^4 - \frac{29}{3} \right]^4. \tag{2.109}$$

In conclusion, I_{OUT} current approximates the Gaussian function using $g_{4b}(x)$ approximation function:

$$I_{\mathrm{OUT}} = I_O g_{4b}\left(\frac{I_{\mathrm{IN}}}{I_O}\right) \cong I_O \exp\left[-\left(\frac{I_{\mathrm{IN}}}{I_O}\right)^2 \right]. \tag{2.110}$$

2.5.2.2 Implementation of the General Gaussian Function

The block diagram of the Gaussian function synthesizer circuit that is able to implement the general form of $g_{4b}(x)$ approximation function is derived of the previous block diagram, realizing the $x \rightarrow x/\sigma\sqrt{2}$ variable changing (Fig. 2.32).

The expressions of I_{Ga}, I_{Gb}, and I_{Gc} currents are

$$I_{\mathrm{Ga}} = \frac{I_{\mathrm{IN}}^2}{32\sigma^2 I_O}, \tag{2.111}$$

$$I_{\mathrm{Gb}} = \frac{32 I_O^2}{3 I_{\mathrm{OUT}'}} = \frac{32}{3} \frac{I_O^2}{I_{G1} + I_O} = \frac{32}{3} \frac{I_O^2}{\frac{I_{\mathrm{IN}}^2}{32\sigma^2 I_O} + I_O} \tag{2.112}$$

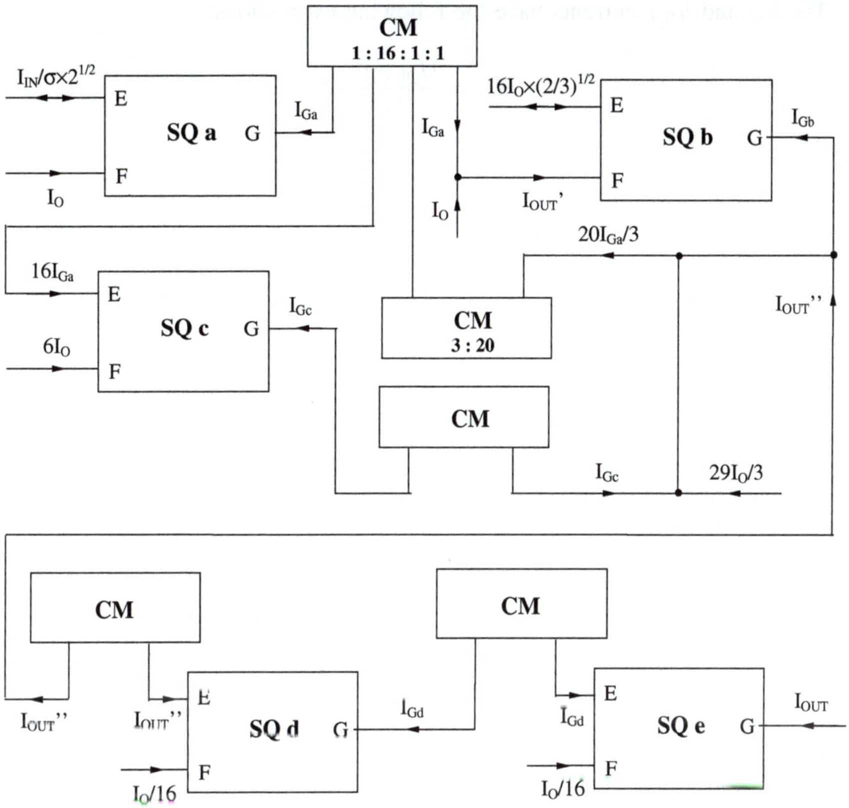

Fig. 2.32 Block diagram of the general Gaussian function synthesizer circuit that uses the $g_{4b}(x)$ approximation function

and

$$I_{\mathrm{Gc}} = \frac{8}{3}\frac{I_{G1}^2}{I_O} = \frac{I_{\mathrm{IN}}^4}{384\sigma^4 I_O^3}. \tag{2.113}$$

The $I_{\mathrm{OUT}''}$ current can be expressed as follows:

$$I_{\mathrm{OUT}''} = I_{\mathrm{Gb}} + \frac{20}{3}I_{\mathrm{Ga}} - I_{\mathrm{Gc}} - \frac{29}{3}I_O, \tag{2.114}$$

resulting

$$I_{\mathrm{OUT}''} = I_O\left[\frac{32}{3}\frac{1}{1+\frac{1}{32\sigma^2}\left(\frac{I_{\mathrm{IN}}}{I_O}\right)^2} + \frac{5}{24\sigma^2}\left(\frac{I_{\mathrm{IN}}}{I_O}\right)^2 - \frac{1}{384\sigma^4}\left(\frac{I_{\mathrm{IN}}}{I_O}\right)^4 - \frac{29}{3}\right]. \tag{2.115}$$

The I_{Gd} and I_{OUT} currents have the following expressions:

$$I_{\mathrm{Gd}} = \frac{\left(I_{\mathrm{OUT}''}\right)^2}{I_O} \tag{2.116}$$

and

$$I_{\mathrm{OUT}} = \frac{I_{Gd}^2}{I_O} = \frac{\left(I_{\mathrm{OUT}''}\right)^4}{I_O^3}. \tag{2.117}$$

Thus, the output current of the circuit can be expressed as follows:

$$I_{\mathrm{OUT}} = I_O\left[\frac{32}{3}\frac{1}{1+\frac{1}{32\sigma^2}\left(\frac{I_{\mathrm{IN}}}{I_O}\right)^2} + \frac{5}{24\sigma^2}\left(\frac{I_{\mathrm{IN}}}{I_O}\right)^2 - \frac{1}{384\sigma^4}\left(\frac{I_{\mathrm{IN}}}{I_O}\right)^4 - \frac{29}{3}\right]^4. \tag{2.118}$$

In conclusion, I_{OUT} current approximates the Gaussian function using $g_{4b}(x)$ approximation function:

$$I_{\mathrm{OUT}} = I_O g_{4b}\left[\frac{1}{\sigma\sqrt{2}}\left(\frac{I_{\mathrm{IN}}}{I_O}\right)\right] \cong I_O \exp\left[-\frac{1}{2\sigma^2}\left(\frac{I_{\mathrm{IN}}}{I_O}\right)^2\right]. \tag{2.119}$$

References

1. Cosmin Popa, "Low-Voltage Improved Accuracy Gaussian Function Generator with Fourth-Order Approximation", Microelectronics Journal, Volume 43, Issue 8, August 2012, pp. 515–520.
2. Cosmin Popa, "High-Accuracy Function Synthesizer Circuit with Applications in Signal Processing", EURASIP Journal on Advanced in Signal Processing, 2012.

Chapter 3
Hyperbolic Functions' Synthesizers

3.1 Introduction

Hyperbolic mathematical functions present a multitude of applications in VLSI designs and analog signal processing structures. As any device characteristic modeling the hyperbolic functions (such as hyperbolic sinusoidal, co-sinusoidal, or tangent functions, for example) does not exist in CMOS technology, the possibility of an accurate generation of these functions is strongly restricted to the utilization of superior-order approximation functions. Particular approximation functions have to be developed, considering the restrictions imposed by their implementation in CMOS technology. The minimization of the computational circuits' complexity is correlated with the development of approximation functions in such a way that their concrete implementations in CMOS technology use the simplest existent fundamental computational blocks. In this specific technological context, the most convenient choices for current-mode fundamental blocks are represented (in this order) by squaring and multiplier/divider structures.

The development of accurate superior-order approximation functions will represent the main part of this chapter, the presented functions being used for designing high-precision hyperbolic sinusoidal, co-sinusoidal, and tangent function generators.

3.2 Synthesis of Hyperbolic Sinusoidal Function (sinh Function)

The hyperbolic sinusoidal function can be generated using specific CMOS computational circuits, by developing accurate superior-order approximation functions. The practical requirement for a good frequency response, associated with the necessity of synthesizing performance structures, having an operation that have to be as small as possible dependent on temperature and technological-cause errors, imposes the design of hyperbolic sinusoidal function synthesizers for a current-mode operation.

C. R. Popa, *Current-Mode Analog Nonlinear Function Synthesizer Structures*,
DOI: 10.1007/978-3-319-01035-9_3,

3.2.1 Approximation of Hyperbolic Sinusoidal Function Using Taylor Series

The $\sinh(x)$ mathematical function can be expressed as follows:

$$\sinh(x) = \frac{e^x - e^{-x}}{2}. \tag{3.1}$$

In order to develop accurate superior-order approximations of hyperbolic sinusoidal function, the expression of its Taylor series expansion must be determined. The fifth-order limited series expansion can be expressed as follows:

$$\sinh(x) = x + \frac{x^3}{6} + \frac{x^5}{120} + \ldots \tag{3.2}$$

As $\sinh(x)$ is an odd function, the even-order terms from its previous expansions are zero.

3.2.2 Third-Order Approximation of Hyperbolic Sinusoidal Function

The method presented for generating the hyperbolic sinusoidal function is based on the utilization of a third-order approximation function, developed considering the previous mentioned particularities of the implemented function. Taking into account the considerations referring to a facile implementation in CMOS technology using minimal hardware resources, a third-order approximation function that can be used for synthesizing the hyperbolic sinusoidal functions has the following particular expression:

$$g_1(x) = -\frac{6}{x + \sqrt{6}} + \frac{x^2}{\sqrt{6}} + \sqrt{6}. \tag{3.3}$$

For relatively small values of the x input variable, the error of approximation is mainly given by the fourth-order term, neglected in the approximation:

$$\varepsilon_{f(x)}^{g_1(x)} \cong \frac{x^4}{6\sqrt{6}\sinh(x)}. \tag{3.4}$$

A comparison between the $f(x) = \sinh(x)$ function and $g_1(x)$ approximation function is shown in Table 3.1.

The graphical representations of $f(x) = \sinh(x)$ and $g_1(x)$ functions are shown in Fig. 3.1, while the graphical representation of the approximation error, $\varepsilon(x)$, defined as the difference between $g_1(x)$ and $f(x) = \sinh(x)$ functions, is presented in Fig. 3.2.

Table 3.1 Comparison between the $f(x) = \sinh(x)$ function and $g_1(x)$ approximation function

x [rad]	$f(x)$	$g_1(x)$	ε
−1.0	−1.18	−1.28	−0.1
−0.8	−0.89	0.93	−0.04
−0.6	−0.64	−0.64	0.00
−0.4	−0.41	−0.41	0.00
−0.2	−0.20	−0.20	0.00
0.0	0.00	0.00	0.00
0.2	0.20	0.20	0.00
0.4	0.41	0.41	0.00
0.6	0.64	0.63	−0.01
0.8	0.89	0.87	−0.02
1.0	1.18	1.12	−0.06
1.2	1.51	1.39	−0.12

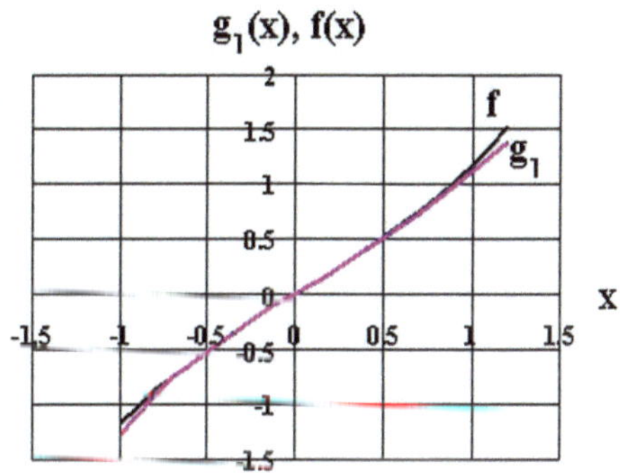

Fig. 3.1 Graphical representation of $f(x) = \sinh(x)$ and $g_1(x)$ functions

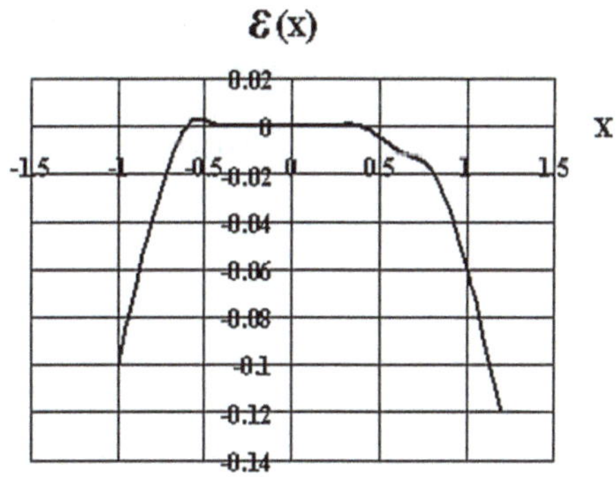

Fig. 3.2 Graphical representation of the approximation error, $\varepsilon(x)$

The block diagram of the hyperbolic sinusoidal function synthesizer circuit based on $g_1(x)$ approximation function is presented in Fig. 3.3.

The "SQ" block from Fig. 3.3 represents a current-mode squaring circuit, having the concrete implementation presented in Fig. 3.4 [1].

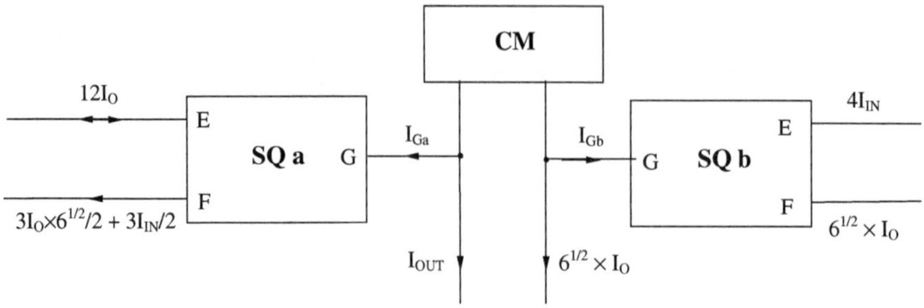

Fig. 3.3 Block diagram of the hyperbolic sinusoidal function synthesizer circuit based on $g_1(x)$ approximation function

Fig. 3.4 Squaring circuit (SQ) implementation [1]

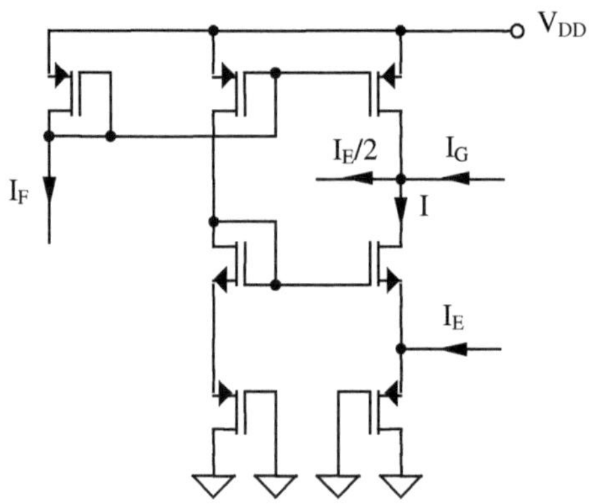

Noting with $V_{\mathrm{GS}}(I)$, the absolute value of the gate–source voltage of a MOS transistor biased at a drain current equal with I, the equation of the translinear loop can be expressed as follows:

$$2V_{\mathrm{GS}}(I_F) = V_{\mathrm{GS}}(I) + V_{\mathrm{GS}}(I + I_E), \tag{3.5}$$

resulting

$$2\sqrt{I_E} = \sqrt{I} + \sqrt{I + I_E}. \tag{3.6}$$

So

$$I = I_F - \frac{I_E}{2} + \frac{I_E^2}{16 I_F}. \tag{3.7}$$

The expression of the output current will be

$$I_G = I + \frac{I_E}{2} - I_F = \frac{I_E^2}{16 I_F}. \tag{3.8}$$

The I_{Ga} and I_{Gb} currents can be expressed as follows:

$$I_{Ga} = \frac{144 I_O^2}{16\left(\frac{3I_O\sqrt{6}}{2} + \frac{3I_{IN}}{2}\right)} \tag{3.9}$$

and

$$I_{Gb} = \frac{16 I_{IN}^2}{16 I_O\sqrt{6}}, \tag{3.10}$$

resulting is the following expression of the output current:

$$I_{OUT} = I_{Gb} - I_{Ga} + I_O\sqrt{6} = -\frac{6I_O^2}{I_{IN} + I_O\sqrt{6}} + \frac{I_{IN}^2}{I_O\sqrt{6}} + I_O\sqrt{6}, \tag{3.11}$$

equivalent to

$$I_{OUT} = I_O\left[-\frac{6}{\left(\frac{I_{IN}}{I_O}\right) + \sqrt{6}} + \frac{1}{\sqrt{6}}\left(\frac{I_{IN}}{I_O}\right)^2 + \sqrt{6}\right]. \tag{3.12}$$

So, I_{OUT} current approximates the hyperbolic sinusoidal function using $g_1(x)$ approximation function:

$$I_{OUT} = I_O g_1\left(\frac{I_{IN}}{I_O}\right) \cong I_O \sinh\left(\frac{I_{IN}}{I_O}\right). \tag{3.13}$$

The CMOS implementation of the hyperbolic sinusoidal function synthesizer circuit based on $g_1(x)$ approximation function is presented in Fig. 3.5 [1].

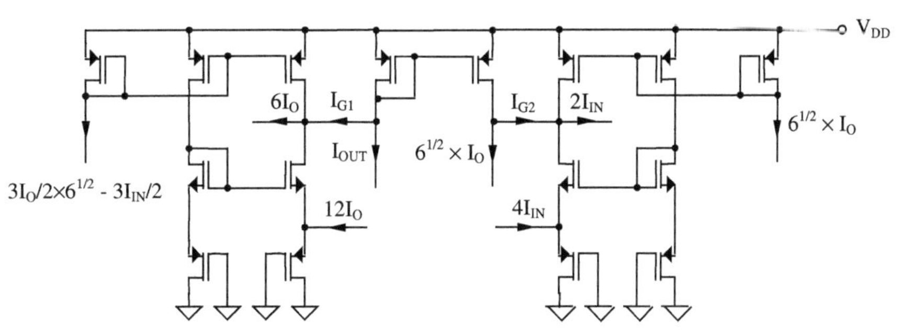

Fig. 3.5 CMOS implementation of the hyperbolic sinusoidal function synthesizer circuit based on $g_1(x)$ approximation function [1]

3.2.3 *Fifth-Order Approximation of Hyperbolic Sinusoidal Function (First Implementation)*

For applications that require a more accurate generation of the hyperbolic sinusoidal function, the utilization of a fifth-order approximation function could be more convenient. A possible form of this function is presented in (3.14). In order to determine its general form, two conditions have been taken into account. First, restrictions referring to its facile implementation in CMOS technology using minimal hardware resources have imposed this general form. Then, the identity between the Taylor series of approximation function and hyperbolic sinusoidal function will determine the concrete values of the approximation function coefficients.

$$g_2(x) = \frac{10\sqrt{5}}{3}\left(\frac{1}{1-\frac{x}{2\sqrt{5}}} - \frac{1}{1+\frac{x}{2\sqrt{5}}}\right) - \frac{7x}{3}. \tag{3.14}$$

A comparison between the $f(x) = \sinh(x)$ function and $g_2(x)$ approximation function is shown in Table 3.2.

Table 3.2 Comparison between the $f(x) = \sinh(x)$ function and $g_2(x)$ approximation function

x [rad]	$f(x)$	$g_2(x)$	ε
−2.4	−5.47	−5.67	−0.20
−2.2	−4.46	−4.56	−0.10
−2.0	−3.63	−3.68	−0.5
−1.8	−2.95	−2.97	−0.02
−1.6	−2.38	−2.39	−0.01
−1.4	−1.91	−1.92	−0.01
−1.2	−1.51	−1.52	−0.01
−1.0	−1.18	−1.18	0.00
−0.8	−0.89	−0.89	0.00
−0.6	−0.64	−0.64	0.00
−0.4	−0.41	−0.41	0.00
−0.2	−0.20	−0.2	0.00
0.0	0.00	0.00	0.00
0.2	0.20	0.20	0.00
0.4	0.41	0.41	0.00
0.6	0.64	0.64	0.00
0.8	0.89	0.89	0.00
1.0	1.18	1.18	0.00
1.2	1.51	1.52	0.01
1.4	1.91	1.92	0.01
1.6	2.38	2.39	0.01
1.8	2.95	2.97	0.02
2.0	3.63	3.68	0.05
2.2	4.46	4.56	0.10
2.4	5.47	5.67	0.20

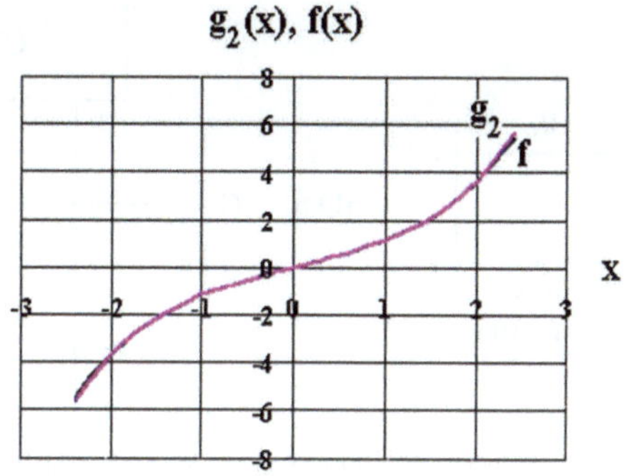

Fig. 3.6 Graphical representation of $f(x) = \sinh(x)$ and $g_2(x)$ functions

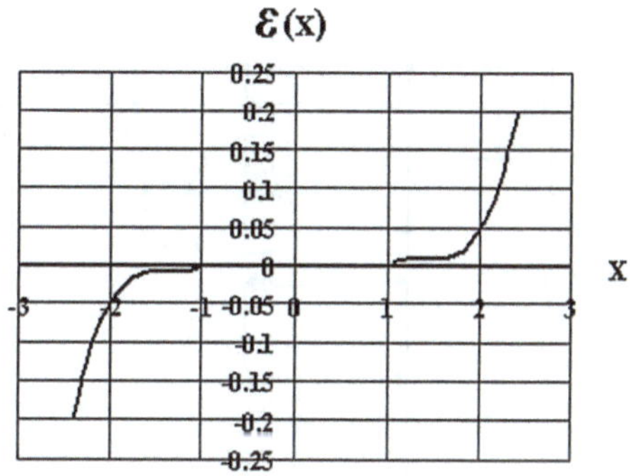

Fig. 3.7 Graphical representation of the approximation error, $\varepsilon(x)$

The graphical representations of $f(x) = \sinh(x)$ and $g_2(x)$ functions are shown in Fig. 3.6, while the graphical representation of the approximation error, $\varepsilon(x)$, defined as the difference between $g_2(x)$ and $f(x) = \sinh(x)$ functions, is presented in Fig. 3.7.

The block diagram of the hyperbolic sinusoidal function synthesizer circuit based on $g_2(x)$ approximation function is presented in Fig. 3.8.

Using the same implementation of the current-mode squaring circuit, the expressions of I_{Ga} and I_{Gb} currents are

$$I_{Ga} = \frac{16 I_O^2}{16\left(\frac{3I_O}{10\sqrt{5}} + \frac{3I_{IN}}{100}\right)}, \tag{3.15}$$

$$I_{Gb} = \frac{16 I_O^2}{16\left(\frac{3I_O}{10\sqrt{5}} - \frac{3I_{IN}}{100}\right)}, \tag{3.16}$$

resulting is the following expression of the output current:

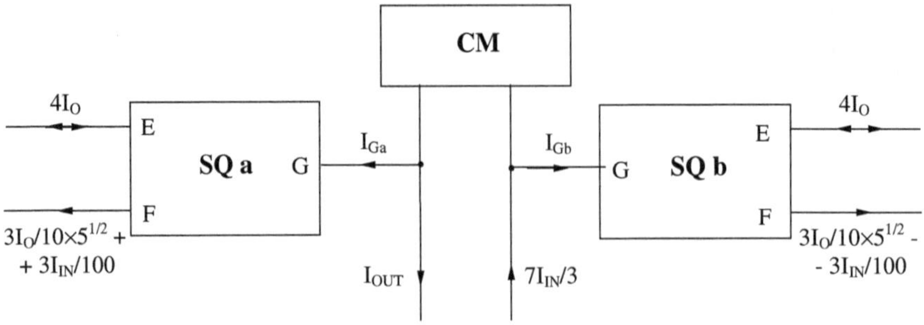

Fig. 3.8 Block diagram of the hyperbolic sinusoidal function synthesizer circuit based on $g_2(x)$ approximation function

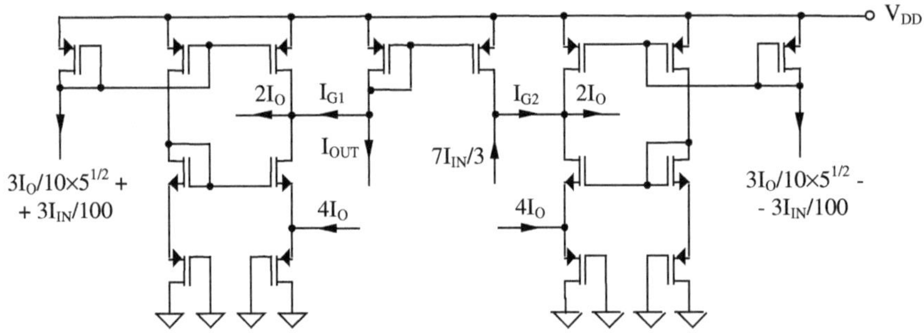

Fig. 3.9 CMOS implementation of the hyperbolic sinusoidal function synthesizer circuit based on $g_2(x)$ approximation function [1]

$$\begin{aligned} I_{\mathrm{OUT}} &= I_{\mathrm{Gb}} - I_{\mathrm{Ga}} - \frac{7I_{\mathrm{IN}}}{3} \\ &= \frac{10\sqrt{5}}{3}\left[\frac{I_O^2}{1-\frac{1}{2\sqrt{5}}\left(\frac{I_{\mathrm{IN}}}{I_O}\right)} - \frac{I_O^2}{1+\frac{1}{2\sqrt{5}}\left(\frac{I_{\mathrm{IN}}}{I_O}\right)}\right] - \frac{7}{3}\left(\frac{I_{\mathrm{IN}}}{I_O}\right) \end{aligned} \tag{3.17}$$

Thus, I_{OUT} current approximates the hyperbolic sinusoidal function using $g_2(x)$ approximation function:

$$I_{\mathrm{OUT}} = I_O g_2\left(\frac{I_{\mathrm{IN}}}{I_O}\right) \cong I_O \sinh\left(\frac{I_{\mathrm{IN}}}{I_O}\right). \tag{3.18}$$

The CMOS implementation of the hyperbolic sinusoidal function synthesizer circuit based on $g_2(x)$ approximation function is presented in Fig. 3.9 [1].

Table 3.3 Comparison between the $f(x) = \sinh(x)$ function and $g_3(x)$ approximation function

x [rad]	$f(x)$	$g_3(x)$	ε
−1.4	−1.91	−1.79	0.12
−1.2	−1.51	−1.45	0.06
−1.0	−1.18	−1.14	0.04
−0.8	−0.89	−0.87	0.02
−0.6	−0.64	−0.63	0.01
−0.4	−0.41	−0.41	0.00
−0.2	−0.20	−0.20	0.00
0.0	0.00	0.00	0.00
0.2	0.20	0.20	0.00
0.4	0.41	0.41	0.00
0.6	0.64	0.64	0.00
0.8	0.89	0.89	0.00
1.0	1.18	1.18	0.00
1.2	1.51	1.50	−0.01
1.4	1.91	1.87	−0.04
1.6	2.38	2.33	−0.05
1.8	2.95	2.85	−0.10
2.0	3.63	3.45	−0.18

3.2.4 Fifth-Order Approximation of Hyperbolic Sinusoidal Function (Second Implementation)

Another possible form of a fifth-order approximation function for generating the hyperbolic sinusoidal function uses two current-mode squaring circuits for the implementation in CMOS technology of its primitive fractional and squaring terms:

$$g_3(x) = \frac{1270}{1 - \frac{x}{20}} \quad 3.166x^2 - 62.5x - 1{,}270. \tag{3.19}$$

The comparison between the $f(x) = \sinh(x)$ function and $g_3(x)$ approximation function is shown in Table 3.3.

The graphical representations of $f(x) = \sinh(x)$ and $g_3(x)$ functions are shown in Fig. 3.10, while the graphical representation of the approximation error, $\varepsilon(x)$, defined as the difference between $g_3(x)$ and $f(x) = \sinh(x)$ functions, is presented in Fig. 3.11.

The hyperbolic sinusoidal function synthesizer circuit based on $g_3(x)$ approximation function has the block diagram presented in Fig. 3.12.

Considering the same realization of the current-mode squaring circuit, the expressions of I_{Ga} and I_{Gb} currents are

$$I_{Ga} = 3.166\frac{I_{IN}^2}{I_O} \tag{3.20}$$

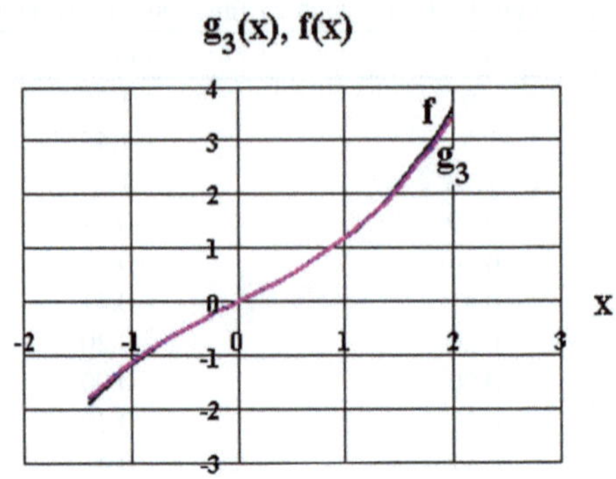

Fig. 3.10 Graphical representation of $f(x) = \sinh(x)$ and $g_3(x)$ functions

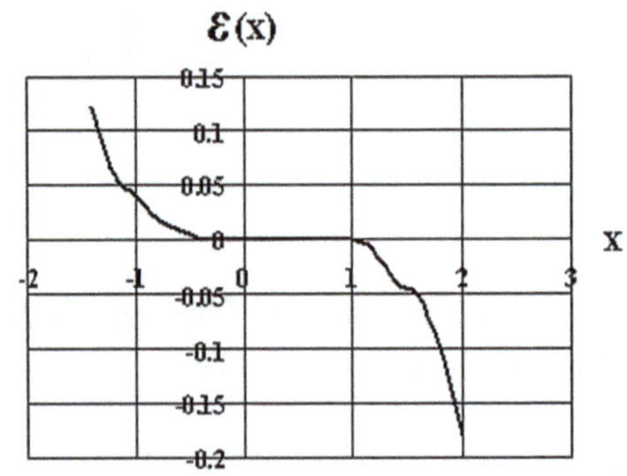

Fig. 3.11 Graphical representation of the approximation error, $\varepsilon(x)$

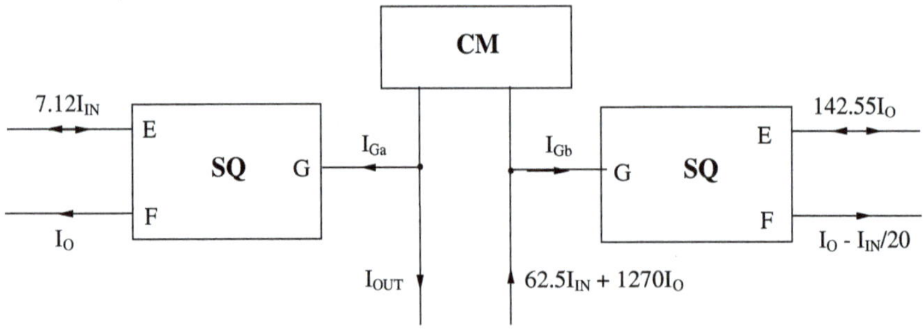

Fig. 3.12 Block diagram of the hyperbolic sinusoidal function synthesizer circuit based on $g_3(x)$ approximation function

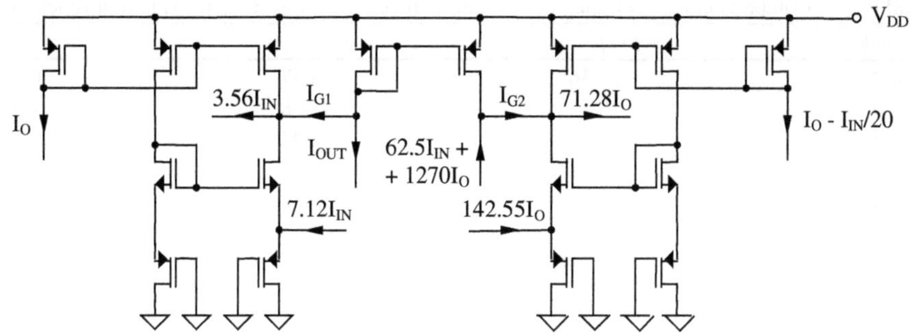

Fig. 3.13 CMOS implementation of the hyperbolic sinusoidal function synthesizer circuit based on $g_3(x)$ approximation function [1]

and

$$I_{Gb} = \frac{1{,}270 I_O^2}{I_O - \frac{I_{IN}}{20}}. \tag{3.21}$$

So, the expression of the output current will be

$$I_{OUT} = I_{Gb} - I_{Ga} - 62.5 I_{IN} - 1{,}270 I_O, \tag{3.22}$$

which is equivalent to

$$I_{OUT} = I_O \left[\frac{1270}{1 - \frac{1}{20}\left(\frac{I_{IN}}{I_O}\right)} - 3.166\left(\frac{I_{IN}}{I_O}\right)^2 - 62.5\left(\frac{I_{IN}}{I_O}\right) - 1{,}270 \right]. \tag{3.23}$$

In conclusion, I_{OUT} current approximates the hyperbolic sinusoidal function using $g_3(x)$ approximation function:

$$I_{OUT} = I_O g_3\left(\frac{I_{IN}}{I_O}\right) \cong I_O \sinh\left(\frac{I_{IN}}{I_O}\right). \tag{3.24}$$

The hyperbolic sinusoidal function synthesizer circuit based on $g_3(x)$ approximation function has the CMOS implementation presented in Fig. 3.13 [1].

3.2.5 Seventh-Order Approximation of Hyperbolic Sinusoidal Function

An additional improvement in the accuracy of generating the hyperbolic sinusoidal function can be obtained by increasing the order of approximation. A seventh-order approximation function having a reasonable complexity of its implementation in CMOS technology can be expressed as follows:

Table 3.4 Comparison between the $f(x) = \sinh(x)$ function and $g_4(x)$ approximation function

x [rad]	$f(x)$	$g_4(x)$	ε
−4.0	−27.36	−28.47	−1.11
−3.6	−18.33	−18.68	−0.35
−3.2	−12.27	−12.37	−0.10
−2.8	−8.21	−8.23	−0.02
−2.4	−5.47	−5.48	−0.01
−2.0	−3.63	−3.63	0.00
−1.6	−2.38	−2.38	0.00
−1.2	−1.51	−1.51	0.00
−0.8	−0.89	−0.89	0.00
−0.4	−0.41	−0.41	0.00
0.0	0.00	0.00	0.00
0.4	0.41	0.41	0.00
0.8	0.89	0.89	0.00
1.2	1.51	1.51	0.00
1.6	2.38	2.38	0.00
2.0	3.63	3.63	0.00
2.4	5.47	5.48	0.01
2.8	8.21	8.23	0.02
3.2	12.27	12.37	0.10
3.6	18.33	18.68	0.35
4.0	27.36	28.48	1.12

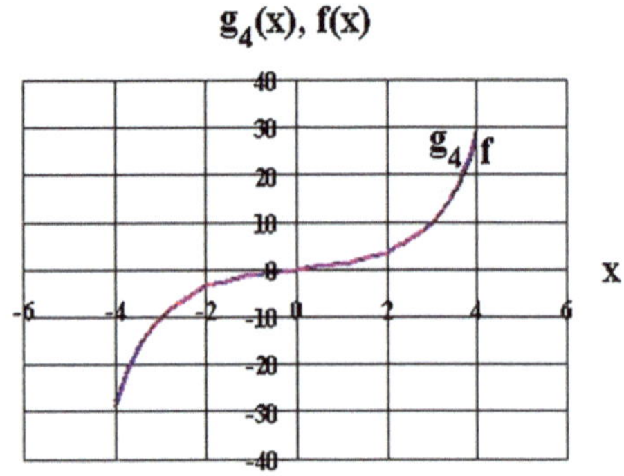

Fig. 3.14 Graphical representation of $f(x) = \sinh(x)$ and $g_4(x)$ functions

$$g_4(x) = \frac{147\sqrt{42}}{20}\left(\frac{1}{1-\frac{x}{\sqrt{42}}} - \frac{1}{1+\frac{x}{\sqrt{42}}}\right) - \frac{11}{60}x^3 - \frac{137}{10}x. \tag{3.25}$$

The comparison between the $f(x) = \sinh(x)$ function and $g_4(x)$ approximation function is shown in Table 3.4.

The graphical representations of $f(x) = \sinh(x)$ and $g_4(x)$ functions are shown in Fig. 3.14, while the graphical representation of the approximation error, $\varepsilon(x)$, defined as the difference between $g_4(x)$ and $f(x) = \sinh(x)$ functions, is presented in Fig. 3.15.

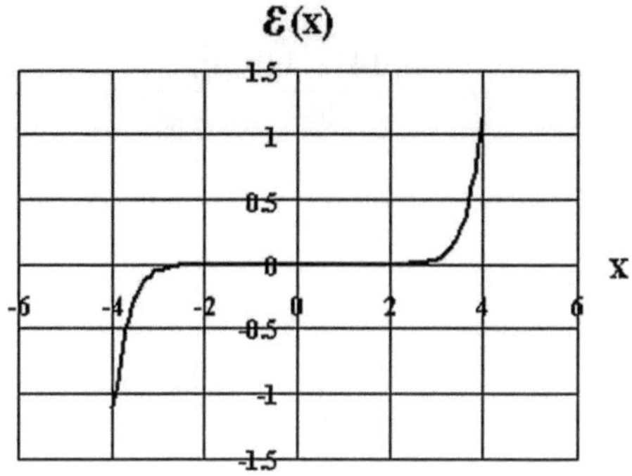

Fig. 3.15 Graphical representation of the approximation error, $\varepsilon(x)$

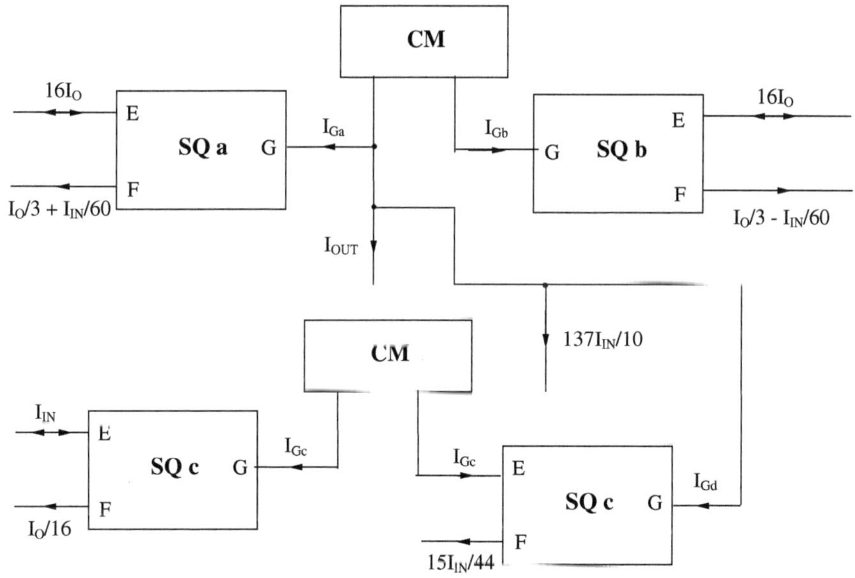

Fig. 3.16 Block diagram of the hyperbolic sinusoidal function synthesizer circuit based on $g_4(x)$ approximation function

The block diagram of the hyperbolic sinusoidal function synthesizer circuit based on $g_4(x)$ approximation function is presented in Fig. 3.16.

The I_{Ga}, I_{Gb}, I_{Gc}, and I_{Gd} currents can be expressed (using the same implementation of the current-mode squaring circuit) as follows:

$$I_{Ga} = \frac{147\sqrt{42}}{20} \frac{I_O^2}{I_O + \frac{I_{IN}}{20}}, \tag{3.26}$$

$$I_{Gb} = \frac{147\sqrt{42}}{20}\frac{I_O^2}{I_O - \frac{I_{IN}}{20}}, \tag{3.27}$$

$$I_{Gc} = \frac{I_{IN}^2}{I_O} \tag{3.28}$$

and

$$I_{Gd} = \frac{11\,I_{G3}^2}{60\,I_{IN}} = \frac{11\,I_{IN}^3}{60\,I_O^2}, \tag{3.29}$$

resulting is the following expression of the output current:

$$I_{OUT} = I_{Gb} - I_{Ga} - I_{Gd} - \frac{1,37 I_{IN}}{10}, \tag{3.30}$$

equivalent to

$$I_{OUT} = I_O\left[\frac{147\sqrt{42}}{20}\left(\frac{1}{1-\frac{1}{\sqrt{42}}\frac{I_{IN}}{I_O}} - \frac{1}{1+\frac{1}{\sqrt{42}}\frac{I_{IN}}{I_O}}\right) - \frac{11}{60}\left(\frac{I_{IN}}{I_O}\right)^3 - \frac{137}{10}\frac{I_{IN}}{I_O}\right]. \tag{3.31}$$

So, I_{OUT} current approximates the hyperbolic sinusoidal function using $g_4(x)$ approximation function:

$$I_{OUT} = I_O g_4\left(\frac{I_{IN}}{I_O}\right) \cong I_O \sinh\left(\frac{I_{IN}}{I_O}\right). \tag{3.32}$$

3.3 Synthesis of Hyperbolic Co-Sinusoidal Function (cosh Function)

The accurate generation of hyperbolic co-sinusoidal function can be also achieved using superior-order approximation functions. Similar to the computational structures that are able to implement the hyperbolic sinusoidal function, the development of the general form of approximation functions for generating the hyperbolic co-sinusoidal function must take into account the same restrictions referring to the complexity of their CMOS implementations and also must fulfilled the most important requirements referring to the accuracy of the approximation and to the circuits' frequency response.

3.3.1 Approximation of Hyperbolic Co-Sinusoidal Function Using Taylor series

The cosh function can be expressed as follows:

$$\cosh(x) = \frac{e^x + e^{-x}}{2}. \tag{3.33}$$

The sixth-order limited series expansion has the following expression:

$$\cosh(x) = 1 + \frac{x^2}{2} + \frac{x^4}{24} + \frac{x^6}{720} + \ldots. \tag{3.34}$$

Because $\cosh(x)$ is an even function, the odd-order terms from its previous expansions are zero.

3.3.2 Fourth-Order Approximation of Hyperbolic Co-Sinusoidal Function

The following fourth-order approximation function for developing the hyperbolic co-sinusoidal function has a general form that can be easily implemented in CMOS technology using only two current-mode squaring circuits:

$$g_5(x) = 3\left(\frac{1}{1 - \frac{x}{2\sqrt{3}}} + \frac{1}{1 + \frac{x}{2\sqrt{3}}}\right) - 5. \tag{3.35}$$

Table 3.5 Comparison between the $f(x) = \cosh(x)$ function and $g_5(x)$ approximation function

x [rad]	$f(x)$	$g_5(x)$	ε
−2.0	3.76	4.00	0.24
−1.8	3.11	3.22	0.11
−1.6	2.58	2.67	0.09
−1.4	2.15	2.17	0.02
−1.2	1.81	1.82	0.01
−1.0	1.54	1.55	0.01
−0.8	1.34	1.34	0.00
−0.6	1.19	1.19	0.00
−0.4	1.08	1.08	0.00
−0.2	1.02	1.02	0.00
0.0	1.0	1.0	0.00
0.2	1.02	1.02	0.00
0.4	1.08	1.08	0.00
0.6	1.19	1.19	0.00
0.8	1.34	1.34	0.00
1.0	1.54	1.55	0.01
1.2	1.81	1.82	0.01
1.4	2.15	2.17	0.02
1.6	2.58	2.67	0.09
1.8	3.11	3.22	0.11
2.0	3.76	4.00	0.24

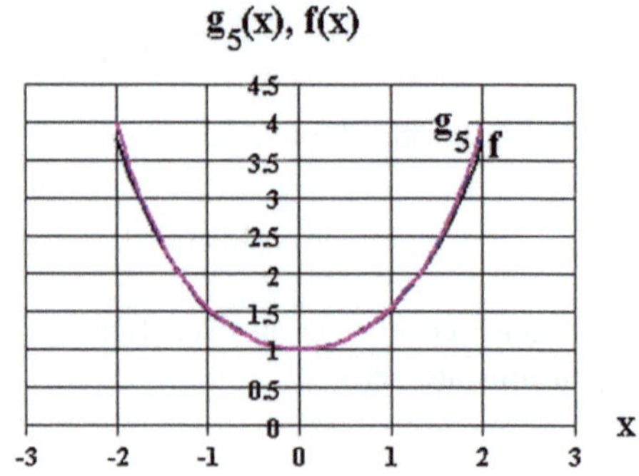

Fig. 3.17 Graphical representation of $f(x) = \cosh(x)$ and $g_5(x)$ functions

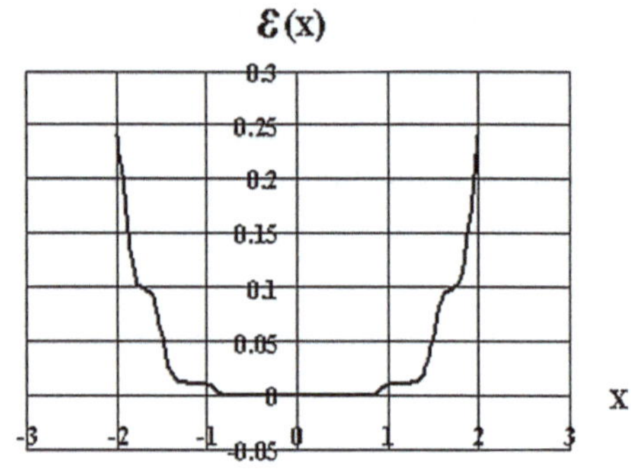

Fig. 3.18 Graphical representation of the approximation error, $\varepsilon(x)$

A comparison between the $f(x) = \cosh(x)$ function and $g_5(x)$ approximation function is shown in Table 3.5.

The graphical representations of $f(x) = \cosh(x)$ and $g_5(x)$ functions are shown in Fig. 3.17, while the graphical representation of the approximation error, $\varepsilon(x)$, defined as the difference between $g_5(x)$ and $f(x) = \cosh(x)$ functions, is presented in Fig. 3.18.

The block diagram of the hyperbolic co-sinusoidal function synthesizer circuit based on $g_5(x)$ approximation function is presented in Fig. 3.19.

The expressions of I_{Ga} and I_{Gb} currents are

$$I_{\mathrm{Ga}} = \frac{16 I_O^2}{16\left(\frac{I_O}{3} - \frac{I_{\mathrm{IN}}}{6\sqrt{3}}\right)}, \tag{3.36}$$

$$I_{\mathrm{Gb}} = \frac{16 I_O^2}{16\left(\frac{I_O}{3} + \frac{I_{\mathrm{IN}}}{6\sqrt{3}}\right)}, \tag{3.37}$$

resulting is the following expression of the output current:

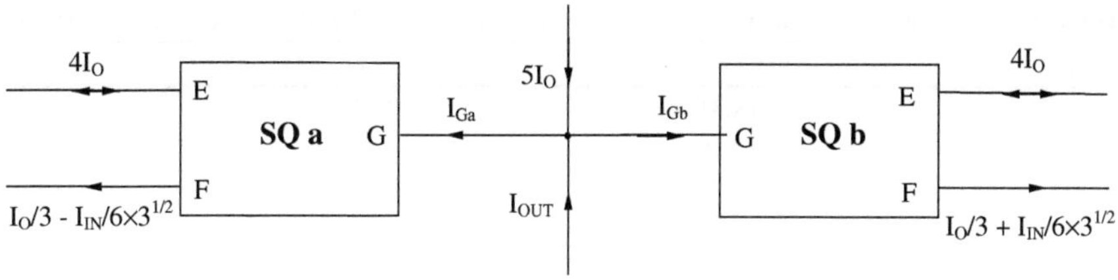

Fig. 3.19 Block diagram of the hyperbolic co-sinusoidal function synthesizer circuit based on $g_5(x)$ approximation function

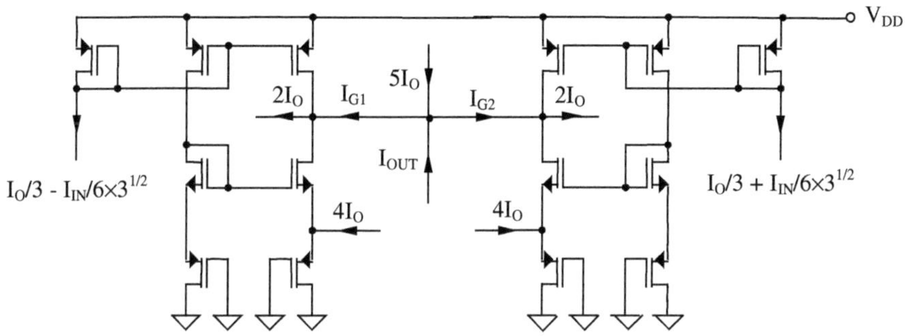

Fig. 3.20 CMOS implementation of the hyperbolic co-sinusoidal function synthesizer circuit based on $g_5(x)$ approximation function [1]

$$I_{OUT} = I_{Ga} + I_{Gb} - 5I_O = \frac{I_O^2}{\frac{I_O}{3} - \frac{I_{IN}}{6\sqrt{3}}} + \frac{I_O^2}{\frac{I_O}{3} + \frac{I_{IN}}{6\sqrt{3}}} - 5I_O, \tag{3.38}$$

equivalent to

$$I_{OUT} = I_O\left[\frac{3}{1 - \frac{1}{2\sqrt{3}}\left(\frac{I_{IN}}{I_O}\right)} + \frac{3}{1 + \frac{1}{2\sqrt{3}}\left(\frac{I_{IN}}{I_O}\right)} - 5\right]. \tag{3.39}$$

So, I_{OUT} current approximates the hyperbolic co-sinusoidal function using $g_5(x)$ approximation function:

$$I_{OUT} = I_O g_5\left(\frac{I_{IN}}{I_O}\right) \cong I_O \cosh\left(\frac{I_{IN}}{I_O}\right). \tag{3.40}$$

The CMOS implementation of the hyperbolic co-sinusoidal function synthesizer circuit based on $g_5(x)$ approximation function is presented in Fig. 3.20 [1].

Table 3.6 Comparison between the $f(x) = \cosh(x)$ function and $g_6(x)$ approximation function

x [rad]	$f(x)$	$g_6(x)$	ε
−3.6	18.31	19.72	1.41
−3.2	12.29	12.70	0.41
−2.8	8.26	8.34	0.08
−2.4	5.56	5.56	0.00
−2.0	3.76	3.74	−0.02
−1.6	2.58	2.55	−0.03
−1.2	1.81	1.79	−0.02
−0.8	1.34	1.32	−0.02
−0.4	1.08	1.07	−0.01
0.0	1.0	1.0	0.00
0.4	1.08	1.07	−0.01
0.8	1.34	1.32	−0.02
1.2	1.81	1.79	−0.02
1.6	2.58	2.55	−0.03
2.0	3.76	3.74	−0.02
2.4	5.56	5.56	0.00
2.8	8.26	8.34	0.08
3.2	12.29	12.70	0.41
3.6	18.31	19.72	1.41

3.3.3 Sixth-Order Approximation of Hyperbolic Co-Sinusoidal Function (First Implementation)

The increase in the order of approximation from four to six will consequently increase the accuracy of the hyperbolic co-sinusoidal function generator that is based on the new sixth-order approximation function:

$$g_6(x) = \frac{75}{4}\left(\frac{1}{1-\frac{x}{\sqrt{30}}} + \frac{1}{1+\frac{x}{\sqrt{30}}}\right) - \frac{3x^2}{4} - \frac{73}{2}. \tag{3.41}$$

The comparison between the $f(x) = \cosh(x)$ function and $g_6(x)$ approximation function is presented in Table 3.6.

The graphical representations of $f(x) = \cosh(x)$ and $g_6(x)$ functions are shown in Fig. 3.21. The graphical representation of the approximation error, $\varepsilon(x)$, defined as the difference between $g_6(x)$ and $f(x) = \cosh(x)$ functions, is shown in Fig. 3.22.

The hyperbolic co-sinusoidal function synthesizer circuit based on $g_6(x)$ approximation function has the block diagram presented in Fig. 3.23.

The expressions of I_{Ga}, I_{Gb} and I_{Gc} currents are

$$I_{Ga} = \frac{75 I_O^2}{4\left(I_O - \frac{I_{IN}}{\sqrt{30}}\right)}, \tag{3.42}$$

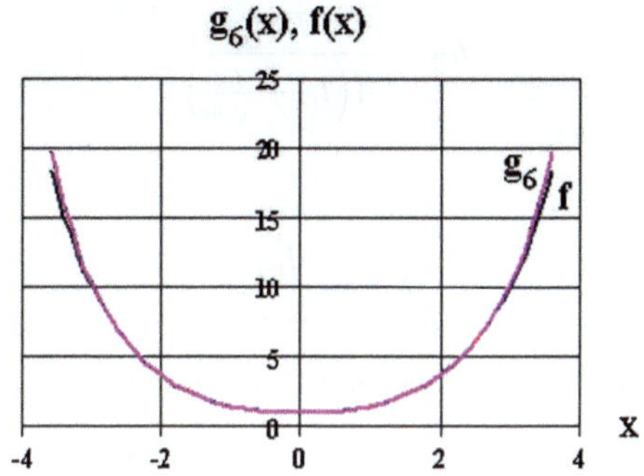

Fig. 3.21 Graphical representation of $g_6(x)$ and $f(x) = \cosh(x)$ functions

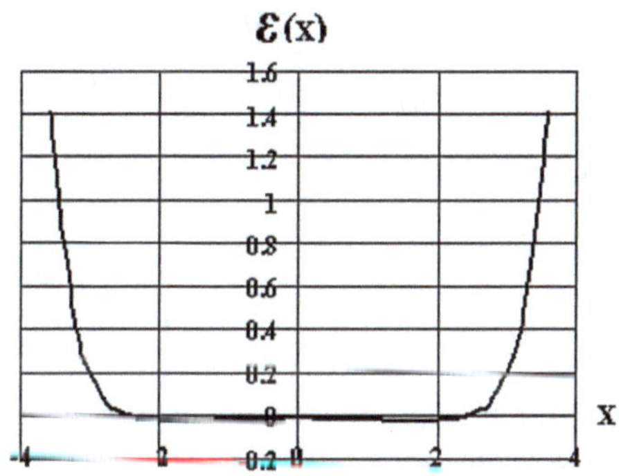

Fig. 3.22 Graphical representation of the approximation error, $\varepsilon(x)$

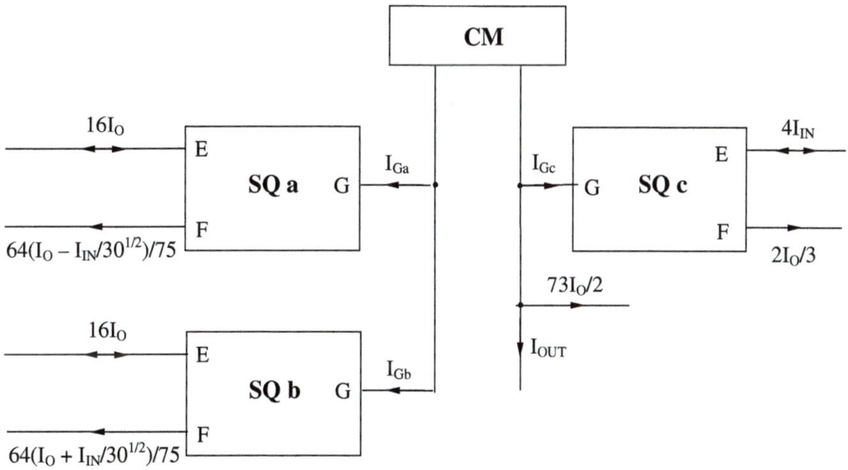

Fig. 3.23 Block diagram of the hyperbolic co-sinusoidal function synthesizer circuit based on $g_6(x)$ approximation function

$$I_{Gb} = \frac{75 I_O^2}{4\left(I_O + \frac{I_{IN}}{\sqrt{30}}\right)} \tag{3.43}$$

and

$$I_{Gc} = \frac{3 I_{IN}^2}{2 I_O}. \tag{3.44}$$

It results in the following expression of the output current:

$$I_{OUT} = I_{Ga} + I_{Gb} - I_{Gc} - \frac{73 I_O}{2}, \tag{3.45}$$

or

$$I_{OUT} = \frac{75 I_O^2}{4\left(I_O - \frac{I_{IN}}{\sqrt{30}}\right)} + \frac{75 I_O^2}{4\left(I_O + \frac{I_{IN}}{\sqrt{30}}\right)} - \frac{3 I_{IN}^2}{2 I_O} - \frac{73 I_O}{2}, \tag{3.46}$$

which is equivalent to

$$I_{OUT} = I_O \left[\frac{75}{4} \frac{1}{1 - \frac{1}{\sqrt{30}}\left(\frac{I_{IN}}{I_O}\right)} + \frac{75}{4} \frac{1}{1 + \frac{1}{\sqrt{30}}\left(\frac{I_{IN}}{I_O}\right)} - \frac{3}{2}\left(\frac{I_{IN}}{I_O}\right) - \frac{73}{2} \right]. \tag{3.47}$$

In conclusion, I_{OUT} current approximates the hyperbolic co-sinusoidal function using $g_6(x)$ approximation function:

$$I_{OUT} = I_O g_6\left(\frac{I_{IN}}{I_O}\right) \cong I_O \cosh\left(\frac{I_{IN}}{I_O}\right). \tag{3.48}$$

3.3.4 Sixth-Order Approximation of Hyperbolic Co-Sinusoidal Function (Second Implementation)

Another form of a sixth-order approximation function, developed for generating the hyperbolic co-sinusoidal function, could be

$$g_7(x) = \frac{75}{2} \frac{1}{1 - \frac{x^2}{30}} - \frac{3x^2}{4} - \frac{73}{2}. \tag{3.49}$$

The particular values of coefficients that appear in the previous function have been obtained from the condition that the Taylor series of the approximation function must match (in a sixth-order approximation) the Taylor series of the hyperbolic co-sinusoidal function. A comparison between the $f(x) = \cosh(x)$ function and $g_6(x)$ approximation function is shown in Table 3.7.

Table 3.7 Comparison between the $f(x) = \cosh(x)$ function and $g_7(x)$ approximation function

x [rad]	$f(x)$	$g_7(x)$	ε
−3.6	18.31	19.80	1.49
−3.2	12.29	12.75	0.46
−2.8	8.26	8.39	0.13
−2.4	5.56	5.59	0.03
−2.0	3.76	3.77	0.01
−1.6	2.58	2.58	0.00
−1.2	1.81	1.81	0.00
−0.8	1.34	1.34	0.00
−0.4	1.08	1.08	0.00
0.0	1.00	1.00	0.00
0.4	1.08	1.08	0.00
0.8	1.34	1.34	0.00
1.2	1.81	1.81	0.00
1.6	2.58	2.58	0.00
2.0	3.76	3.77	0.00
2.4	5.56	5.59	0.03
2.8	8.26	8.39	0.13
3.2	12.29	12.75	0.46
3.6	18.31	19.80	1.49

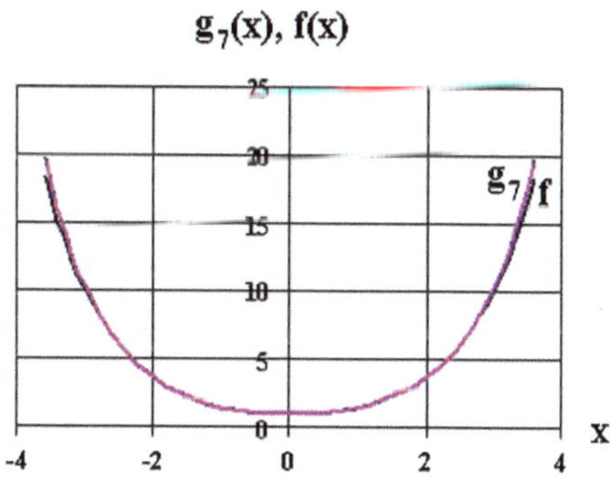

Fig. 3.24 Graphical representation of $g_7(x)$ and $f(x) = \cosh(x)$ functions

The graphical representations of $f(x) = \cosh(x)$ and $g_7(x)$ functions are shown in Fig. 3.24, while the graphical representation of the approximation error, $\varepsilon(x)$, defined as the difference between $g_7(x)$ and $f(x) = \cosh(x)$ functions, is presented in Fig. 3.25.

The block diagram of the hyperbolic co-sinusoidal function synthesizer circuit based on $g_7(x)$ approximation function is presented in Fig. 3.26.

The expressions of I_{Ga}, I and I_{Gb} currents are

$$I_{Ga} = \frac{3}{4}\frac{I_{IN}^2}{I_O}, \tag{3.50}$$

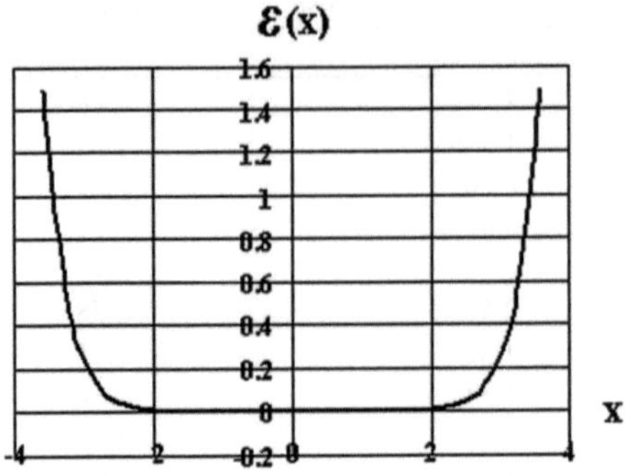

Fig. 3.25 Graphical representation of the approximation error, $\varepsilon(x)$

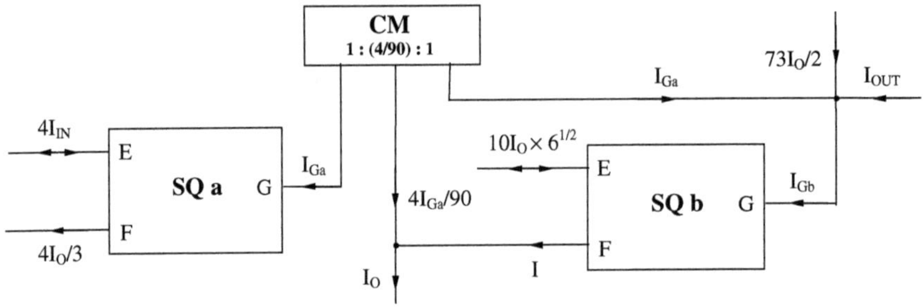

Fig. 3.26 Block diagram of the hyperbolic co-sinusoidal function synthesizer circuit based on $g_7(x)$ approximation function

$$I = I_O - \frac{1}{30}\frac{I_{IN}^2}{I_O}, \tag{3.51}$$

$$I_{Gb} = \frac{600}{16}\frac{I_O^2}{I} = \frac{75}{2}\frac{I_O^2}{I_O - \frac{1}{30}\frac{I_O^2}{I_{IN}}}, \tag{3.52}$$

resulting is the following expression of the output current

$$I_{OUT} = I_{Gb} - I_{Ga} - \frac{73I_O}{2}, \tag{3.53}$$

or

$$I_{OUT} = \frac{75}{2}\frac{I_O^2}{I_O - \frac{1}{30}\frac{I_O^2}{I_{IN}}} - \frac{3}{4}\frac{I_O^2}{I_{IN}} - \frac{73I_O}{2}, \tag{3.54}$$

equivalent to

$$I_{OUT} = I_O\left[\frac{75}{2}\frac{1}{1-\frac{1}{30}\left(\frac{I_{IN}}{I_O}\right)^2} - \frac{3}{4}\left(\frac{I_{IN}}{I_O}\right)^2 - \frac{73}{2}\right]. \tag{3.55}$$

Thus, I_{OUT} current approximates the hyperbolic co-sinusoidal function using $g_7(x)$ approximation function:

$$I_{OUT} = I_O g_7\left(\frac{I_{IN}}{I_O}\right) \cong I_O \cosh\left(\frac{I_{IN}}{I_O}\right). \tag{3.56}$$

3.3.5 Eighth-Order Approximation of Hyperbolic Co-Sinusoidal Function

For applications that require an extreme accuracy of the hyperbolic co-sinusoidal function generation, the following eighth-order approximation function represents a possible solution. Considering the excellent accuracy that could be obtained using this eighth order of approximation, the utilization of four current-mode squaring circuits represents the basis for achieving a reasonable overall circuit complexity.

$$g_8(x) = \frac{5{,}488}{45}\left(\frac{1}{1-\frac{x}{\sqrt{56}}} + \frac{1}{1+\frac{x}{\sqrt{56}}}\right) - \frac{13}{360}x^4 - \frac{347}{90}x^2 - \frac{10931}{45}. \tag{3.57}$$

The comparison between the $f(x) = \cosh(x)$ function and $g_8(x)$ approximation function is shown in Table 3.8.

The graphical representations of $f(x) = \cosh(x)$ and $g_8(x)$ functions are shown in Fig. 3.27, while the graphical representation of the approximation error, $\varepsilon(x)$, defined as the difference between $g_8(x)$ and $f(x) = \cosh(x)$ functions, is presented in Fig. 3.28.

The block diagram of the hyperbolic co-sinusoidal function synthesizer circuit based on $g_8(x)$ approximation function is presented in Fig. 3.29.

The expressions of I_{Ga}, I_{Gb}, I_{Gc} and I_{Gd} currents are

$$I_{Ga} = \frac{5{,}488 I_O^2}{45\left(I_O - \frac{I_{IN}}{\sqrt{56}}\right)}, \tag{3.58}$$

$$I_{Gb} = \frac{5{,}488 I_O^2}{45\left(I_O + \frac{I_{IN}}{\sqrt{56}}\right)}, \tag{3.59}$$

Table 3.8 Comparison between the $f(x) = \cosh(x)$ function and $g_8(x)$ approximation function

x [rad]	$f(x)$	$g_8(x)$	ε
−4.8	60.76	63.76	3.00
−4.4	40.73	41.87	1.14
−4.0	27.31	27.75	0.44
−3.6	18.31	18.50	0.19
−3.2	12.29	12.37	0.08
−2.8	8.26	8.30	0.04
−2.4	5.56	5.59	0.03
−2.0	3.76	3.78	0.02
−1.6	2.58	2.59	0.01
−1.2	1.81	1.82	0.01
−0.8	1.34	1.34	0.00
−0.4	1.08	1.08	0.00
0.0	1.00	1.00	0.00
0.4	1.08	1.08	0.00
0.8	1.34	1.34	0.00
1.2	1.81	1.82	0.01
1.6	2.58	2.59	0.01
2.0	3.76	3.78	0.02
2.4	5.56	5.59	0.03
2.8	8.26	8.30	0.04
3.2	12.29	12.37	0.08
3.6	18.31	18.50	0.19
4.0	27.31	27.75	0.44
4.4	40.73	41.87	1.14
4.8	60.76	63.76	3.00

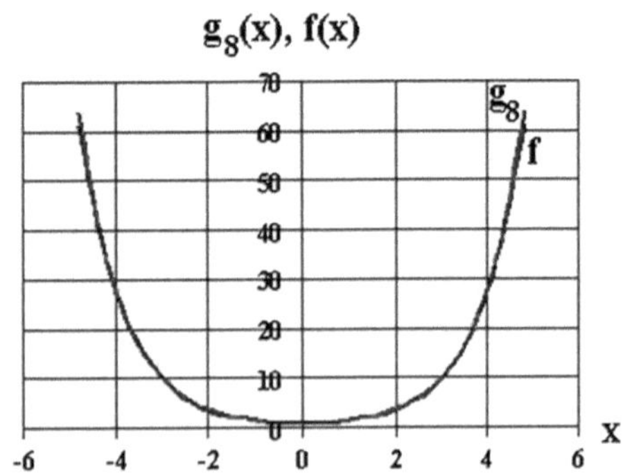

Fig. 3.27 Graphical representation of $g_8(x)$ and $f(x) = \cosh(x)$ functions

$$I_{Gc} = \frac{347 I_{IN}^2}{90 I_O} \tag{3.60}$$

and

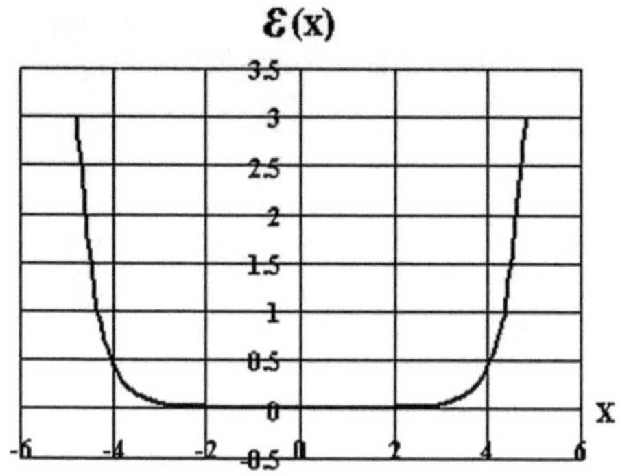

Fig. 3.28 Graphical representation of the approximation error, $\varepsilon(x)$

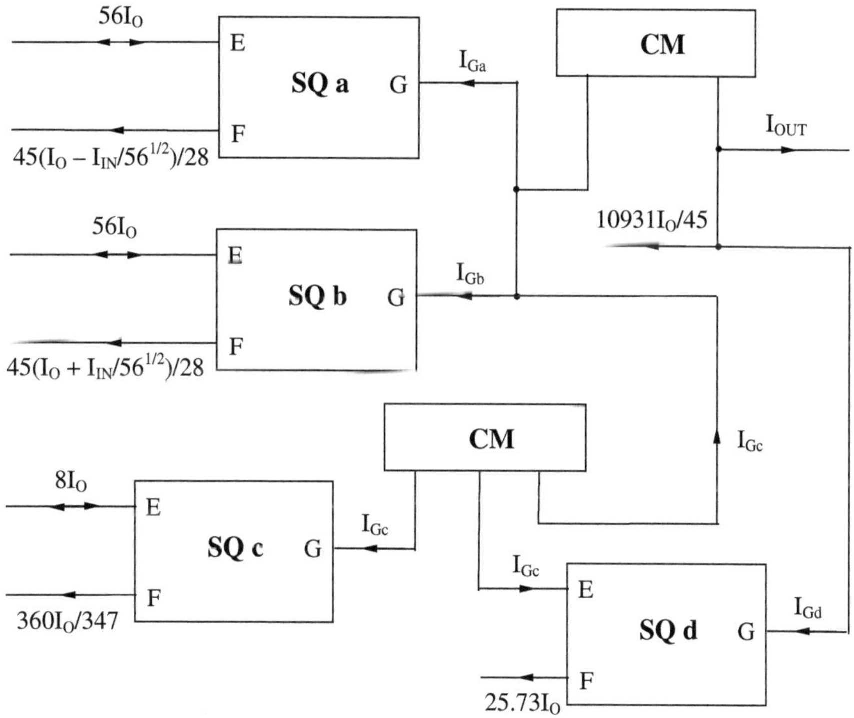

Fig. 3.29 Block diagram of the hyperbolic co-sinusoidal function synthesizer circuit based on $g_8(x)$ approximation function

$$I_{Gd} = \frac{13 I_{IN}^4}{360 I_O^3}. \tag{3.61}$$

It results the following expression of the output current

$$I_{OUT} = I_{Ga} + I_{Gb} - I_{Gc} - I_{Gd} - \frac{10,931}{45} I_O, \tag{3.62}$$

or

$$I_{OUT} = \frac{5,488}{45} \left(\frac{I_O^2}{I_O - \frac{I_{IN}}{\sqrt{56}}} + \frac{I_O^2}{I_O + \frac{I_{IN}}{\sqrt{56}}} \right) - \frac{347 I_{IN}^2}{90 I_O} - \frac{13 I_{IN}^4}{360 I_O^3} - \frac{10,931}{45} I_O, \tag{3.63}$$

equivalent to

$$\begin{aligned} I_{OUT} &= I_O \left[\frac{5,488}{45} \frac{1}{1 - \frac{1}{\sqrt{56}} \left(\frac{I_{IN}}{I_O} \right)} + \frac{5,488}{45} \frac{1}{1 + \frac{1}{\sqrt{56}} \left(\frac{I_{IN}}{I_O} \right)} \right] \\ &- I_O \left[\frac{347}{90} \left(\frac{I_{IN}}{I_O} \right)^2 + \frac{13}{360} \left(\frac{I_{IN}}{I_O} \right)^4 + \frac{10,931}{45} \right] \end{aligned}. \tag{3.64}$$

In conclusion, I_{OUT} current approximates the hyperbolic co-sinusoidal function using $g_8(x)$ approximation function:

$$I_{OUT} = I_O g_8 \left(\frac{I_{IN}}{I_O} \right) \cong I_O \cosh \left(\frac{I_{IN}}{I_O} \right). \tag{3.65}$$

3.4 Synthesis of Hyperbolic Tangent Function (tanh Function)

The generation of accurate hyperbolic tangent function can be obtained using proper superior-order approximation functions. The particular forms of these functions can be obtained taking into account the restrictions imposed by the particularity of approximating an odd function and, also, by the previously mentioned accuracy and technological requirements.

3.4.1 Approximation of Hyperbolic Tangent Function Using Taylor Series

The $\tanh(x)$ function can be expressed as follows:

$$\tanh(x) = \frac{e^x - e^{-x}}{e^x + e^{-x}}. \tag{3.66}$$

In order to develop accurate superior-order approximations of hyperbolic tangent function, the expression of its Taylor series expansion must be determined. The fifth-order limited series expansion of tanh(x) can be expressed as follows:

$$\tanh(x) = x - \frac{x^3}{3} + \frac{2x^5}{5} - \ldots. \tag{3.67}$$

Because tanh(x) is an odd function, the even-order terms from its previous expansions are zero.

3.4.2 Third-Order Approximation of Hyperbolic Tangent Function

A simple third-order approximation function for generating the hyperbolic tangent function with a minimal hardware resource consumption has the following expression:

$$g_9(x) = \frac{1}{1 + \frac{x}{\sqrt[3]{6}}} - \frac{1}{1 - \frac{x}{\sqrt[3]{6}}} + x\left(\frac{2}{\sqrt[3]{6}} + 1\right). \tag{3.68}$$

The particular values of the previous function' coefficients are imposed by the identity between the linear and the third-order terms from the Taylor series expansions of $f(x) = \tanh(x)$ and $g_9(x)$. A comparison between the $f(x) = \tanh(x)$ function and $g_9(x)$ approximation function is shown in Table 3.9.

The $f(x) = \tanh(x)$ and $g_9(x)$ functions have the graphical representations shown in Fig. 3.30. The graphical representation of the approximation error, $\varepsilon(x)$, defined as the difference between $g_9(x)$ and $f(x) = \tanh(x)$ functions, is presented in Fig. 3.31.

The block diagram of the hyperbolic tangent function synthesizer circuit based on $g_9(x)$ approximation function is presented in Fig. 3.32.

The I_{Ga} and I_{Gb} currents can be expressed as follows:

Table 3.9 Comparison between the $f(x) = \tanh(x)$ function and $g_9(x)$ approximation function

x [rad]	$f(x)$	$g_9(x)$	ε
−0.8	−0.66	−0.59	0.07
−0.6	−0.54	−0.52	0.02
−0.4	−0.38	−0.38	0.00
−0.2	−0.20	−0.20	0.00
0.0	0.00	0.00	0.00
0.2	0.20	0.20	0.00
0.4	0.38	0.38	0.00
0.6	0.54	0.52	−0.02
0.8	0.66	0.59	−0.07

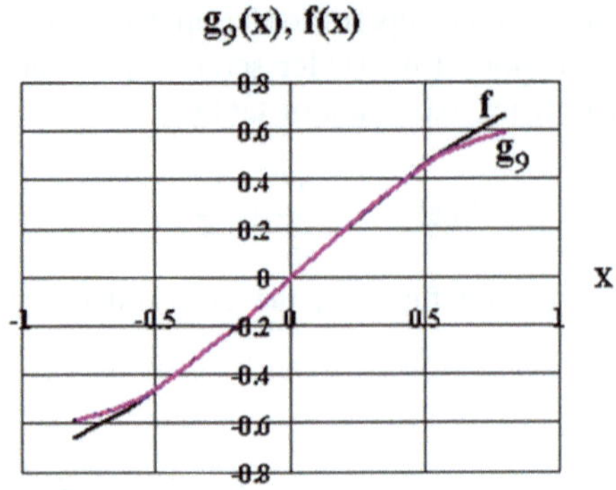

Fig. 3.30 Graphical representation of $f(x) = \tanh(x)$ and $g_9(x)$ functions

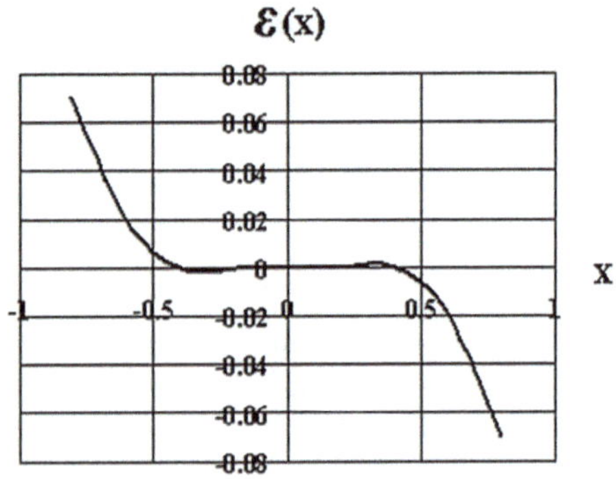

Fig. 3.31 Graphical representation of the approximation error, $\varepsilon(x)$

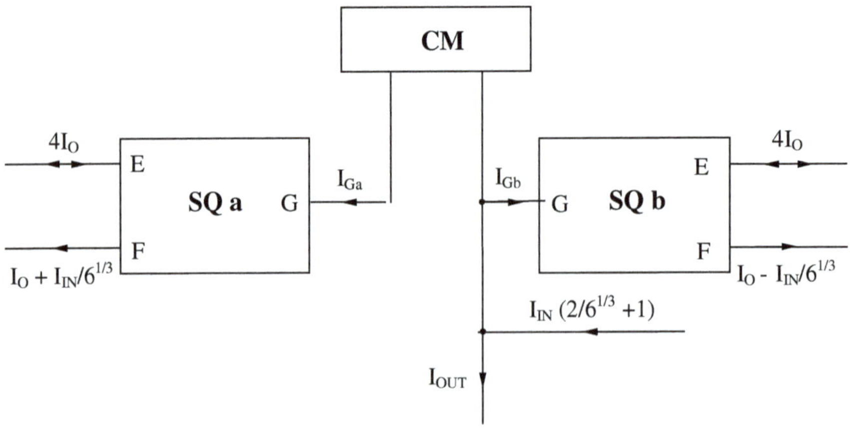

Fig. 3.32 Block diagram of the hyperbolic tangent function synthesizer circuit based on $g_9(x)$ approximation function

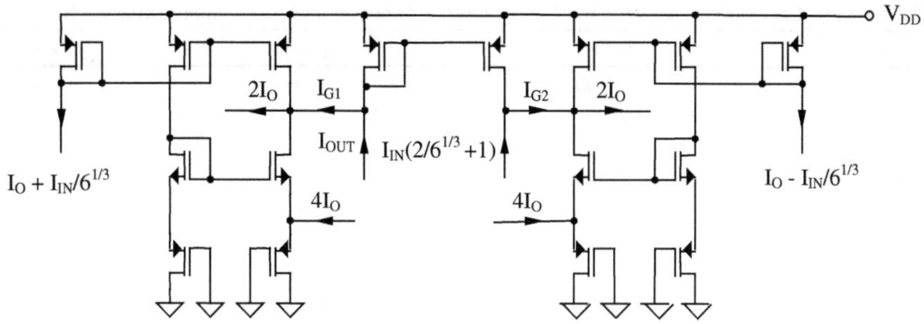

Fig. 3.33 CMOS implementation of the hyperbolic tangent function synthesizer circuit based on $g_9(x)$ approximation function [1]

$$I_{\mathrm{Ga}} = \frac{16 I_O^2}{16\left(I_O + \frac{I_{\mathrm{IN}}}{\sqrt[3]{6}}\right)} \tag{3.69}$$

and

$$I_{\mathrm{Gb}} = \frac{16 I_O^2}{16\left(I_O - \frac{I_{\mathrm{IN}}}{\sqrt[3]{6}}\right)}, \tag{3.70}$$

resulting is the following expression of the output current:

$$I_{\mathrm{OUT}} = I_{\mathrm{Ga}} - I_{\mathrm{Gb}} + \left(\frac{2}{\sqrt[3]{6}} + 1\right) I_{\mathrm{IN}} = \frac{I_O^2}{I_O + \frac{I_{\mathrm{IN}}}{\sqrt[3]{6}}} - \frac{I_O^2}{I_O - \frac{I_{\mathrm{IN}}}{\sqrt[3]{6}}} + \left(\frac{2}{\sqrt[3]{6}} + 1\right) I_{\mathrm{IN}}, \tag{3.71}$$

which is equivalent to

$$I_{\mathrm{OUT}} = I_O \left[\frac{1}{1 + \frac{1}{\sqrt[3]{6}}\left(\frac{I_{\mathrm{IN}}}{I_O}\right)} - \frac{1}{1 - \frac{1}{\sqrt[3]{6}}\left(\frac{I_{\mathrm{IN}}}{I_O}\right)} + \left(\frac{2}{\sqrt[3]{6}} + 1\right)\left(\frac{I_{\mathrm{IN}}}{I_O}\right)\right]. \tag{3.72}$$

Thus, I_{OUT} current approximates the hyperbolic tangent function using $g_9(x)$ approximation function:

$$I_{\mathrm{OUT}} = I_O g_9\left(\frac{I_{\mathrm{IN}}}{I_O}\right) \cong I_O \tanh\left(\frac{I_{\mathrm{IN}}}{I_O}\right). \tag{3.73}$$

The CMOS implementation of the hyperbolic tangent function synthesizer circuit based on $g_9(x)$ approximation function is presented in Fig. 3.33 [1].

Table 3.10 Comparison between the $f(x) = \tanh(x)$ function and $g_{10}(x)$ approximation function

x [rad]	$f(x)$	$g_{10}(x)$	ε
−3.2	−1.00	−1.06	−0.06
−3.0	−0.99	−1.04	−0.05
−2.8	−0.99	−1.03	−0.04
−2.6	−0.99	−1.02	−0.03
−2.4	−0.98	−1.00	−0.02
−2.2	−0.97	−0.99	−0.02
−2.0	−0.96	−0.97	−0.01
−1.8	−0.95	−0.95	0.00
−1.6	−0.92	−0.92	0.00
−1.4	−0.89	−0.89	0.00
−1.2	−0.83	−0.83	0.00
−1.0	−0.76	−0.76	0.00
−0.8	−0.66	−0.66	0.00
−0.6	−0.54	−0.54	0.00
−0.4	−0.38	−0.38	0.00
−0.2	−0.20	−0.20	0.00
0.0	0.00	0.00	0.00
0.2	0.20	0.20	0.00
0.4	0.38	0.38	0.00
0.6	0.54	0.54	0.00
0.8	0.66	0.66	0.00
1.0	0.76	0.76	0.00
1.2	0.83	0.83	0.00
1.4	0.89	0.89	0.00
1.6	0.92	0.92	0.00
1.8	0.95	0.95	0.00
2.0	0.96	0.97	0.01
2.2	0.97	0.99	0.02
2.4	0.98	1.00	0.02
2.6	0.99	1.02	0.03
2.8	0.99	1.03	0.04
3.0	0.99	1.04	0.05
3.2	1.00	1.06	0.06

3.4.3 Fifth-Order Approximation of Hyperbolic Tangent Function

The improvement in the hyperbolic tangent function generation' accuracy can be achieved by increasing the order of approximation. A fifth-order approximation function that can be used for obtaining the required mathematical function can be expressed as follows:

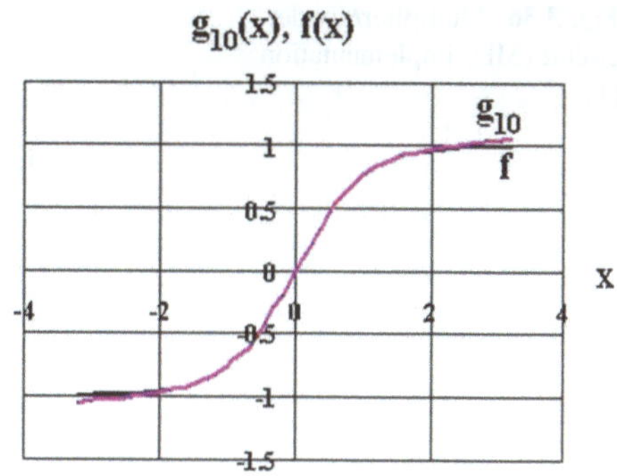

Fig. 3.34 Graphical representation of $f(x) = \tanh(x)$ and $g_{10}(x)$ functions

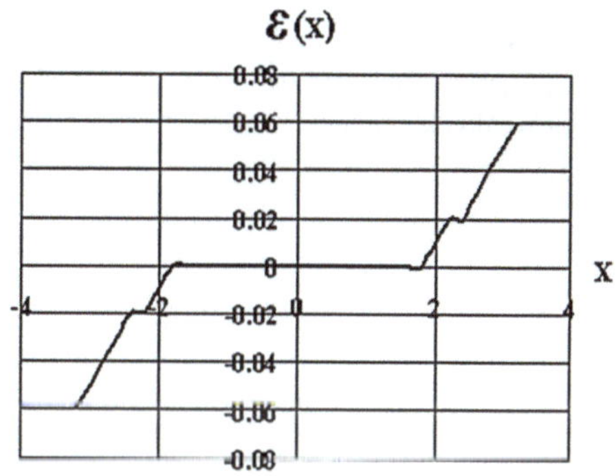

Fig. 3.35 Graphical representation of the approximation error, $\varepsilon(x)$

$$g_{10}(x) = \frac{5}{6}\frac{x}{1 + \frac{2x^2}{5}} + \frac{x}{6}. \tag{3.74}$$

The comparison between the $f(x) = \tanh(x)$ function and $g_{10}(x)$ approximation function is shown in Table 3.10.

The graphical representations of $f(x) = \tanh(x)$ and $g_{10}(x)$ functions are shown in Fig. 3.34, while the graphical representation of the approximation error, $\varepsilon(x)$, defined as the difference between $g_{10}(x)$ and $f(x) = \tanh(x)$ functions, is presented in Fig. 3.35.

The CMOS realization of a current-mode multiplier/divider circuit (MD), required for the implementation of relation (3.74), is presented in Fig. 3.36 [1]. The symbolic representation of the multiplier/divider circuit is shown in Fig. 3.37.

The output current has the following expression:

$$I_D = I_A \frac{I_B}{I_C}. \tag{3.75}$$

The block diagram of the hyperbolic tangent function synthesizer circuit based on $g_{10}(x)$ approximation function is presented in Fig. 3.38.

The expressions of I_G and I currents are

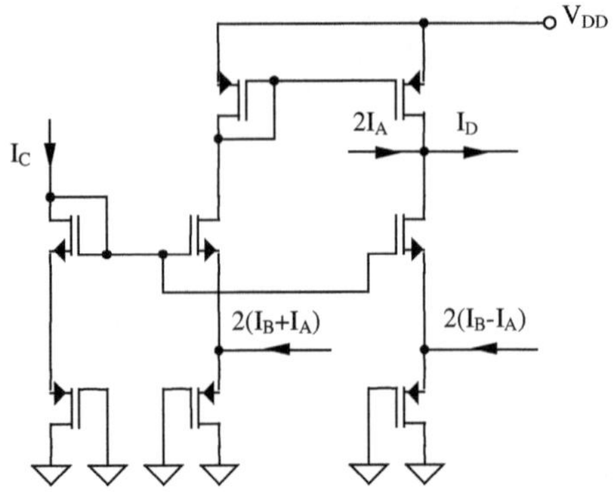

Fig. 3.36 Multiplier/divider circuit (MD) implementation [1]

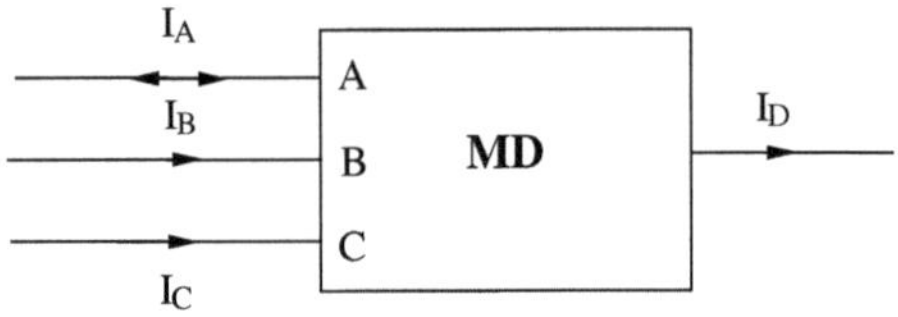

Fig. 3.37 Symbolic representation of the multiplier/divider circuit (MD)

$$I_G = \frac{2}{5}\frac{I_{\rm IN}^2}{I_O} \tag{3.76}$$

and

$$I = \frac{2}{5}\frac{I_{\rm IN}^2}{I_O} + I_O, \tag{3.77}$$

while $I_D$current can be expressed as follows:

$$I_D = \frac{5}{6} I_O \frac{I_{\rm IN}}{I_O + \frac{2}{5}\frac{I_{\rm IN}^2}{I_O}}. \tag{3.78}$$

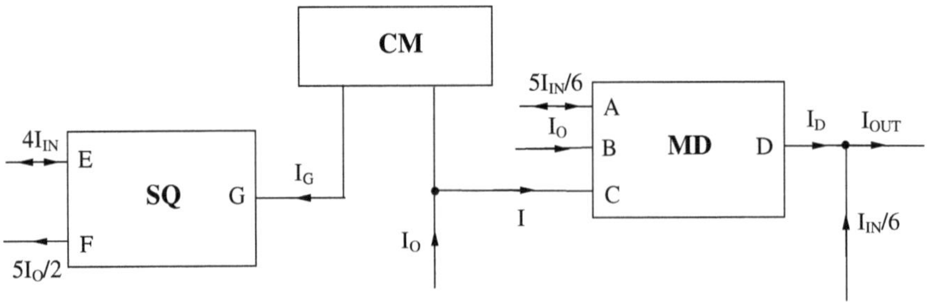

Fig. 3.38 Block diagram of the hyperbolic tangent function synthesizer circuit based on $g_{10}(x)$ approximation function

It results the following expression of the output current:

$$I_{OUT} = I_D + \frac{I_{IN}}{6} = \frac{5}{6} I_O \frac{I_{IN}}{I_O + \frac{2}{5}\frac{I_{IN}^2}{I_O}} + \frac{I_{IN}}{6}, \tag{3.79}$$

which is equivalent to

$$I_{OUT} = I_O \left[\frac{5}{6} \frac{\left(\frac{I_{IN}}{I_O}\right)}{1 + \frac{2}{5}\left(\frac{I_{IN}}{I_O}\right)^2} + \frac{1}{6}\left(\frac{I_{IN}}{I_O}\right) \right]. \tag{3.80}$$

So, I_{OUT} current approximates the hyperbolic tangent function using $g_{10}(x)$ approximation function:

$$I_{OUT} = I_O g_{10}\left(\frac{I_{IN}}{I_O}\right) \cong I_O \tanh\left(\frac{I_{IN}}{I_O}\right). \tag{3.81}$$

Reference

1. Cosmin Popa, "High-Accuracy Function Synthesizer Circuit with Applications in Signal Processing", EURASIP Journal on Advanced in Signal Processing, 2012.

Chapter 4
Third-Order Function Synthesizers

4.1 Introduction

The generation of continuous functions represents an important objective of VLSI integrated circuits developed for analog signal applications. The accuracy of generating these functions is mainly correlated with the complexity of computational circuits that implement function synthesizer structures. From this point of view, the possibility of reducing the overall complexity of analog function synthesizers is strongly related to the modality of implementing the functional core of the computational structure. Classical designs exploit the possibility of generating any continuous function using its superior-order limited Taylor series expansion. This method involves relatively large silicon area consuming, as it requires the individual implementation of each polynomial series term of the expansion. A more convenient method, especially from the perspective of implementing the resulted analog function synthesizer circuits in CMOS technology, is represented by the utilization of superior-order approximation functions, which are able to approximate with very good accuracy the required circuit function. The utilization of this method imposes the consideration of some design restrictions. The most important is related to the accuracy of functions' generation, the increasing of this order certainly increasing the complexity of the function synthesizer CMOS implementation. From this point of view, the development of the superior-order approximation functions using particular primitive functions represented by fractional and squaring functions (having the simplest implementations in CMOS technology) is usually preferred. The additional utilization of the current-mode operation for basic circuits that implement these primitive functions will increase the accuracy of function synthesizer structures and, also, will improve their frequency response. The low-voltage operation of analog function synthesizers using superior-order approximation functions can be achieved using specific design techniques, such as current-mode operation.

The second-order approximation is not able to match with enough accuracy a continuous function, resulting the necessity of increasing the order of approximation for improving the analog function synthesizer' performances. In the

C. R. Popa, *Current-Mode Analog Nonlinear Function Synthesizer Structures*,
DOI: 10.1007/978-3-319-01035-9_4,

context of the equilibrium that must be considered between the accuracy of the computational circuit and the complexity of its implementation in CMOS technology, the third-order of the approximation represents, for most of continuous functions, a convenient choice. The third-order match between $f(x)$ approximated function and $g(x)$ approximation function is equivalent with the identity of the first four terms from their expansions: the constant term, the linear term, the second-order, and the third-order term. In order to reduce the complexity of the CMOS circuits that implement the following presented approximation functions, they will be developed using only two primitive functions and two additional terms. These terms will not require any supplementary computational circuits, as they can be implemented using linear relations between the currents from the circuit. The third-order Taylor series of $f(x)$ function can be considered to have the following general expression:

$$f(x) \cong m + nx + px^2 + qx^3, \tag{4.1}$$

m, n, p, and q being the coefficients of the expansions, determined by the particular expression of $f(x)$ function.

4.2 Primitive Continuous Functions

The following presented method for generating a multitude of $f(x)$ continuous circuit functions using a $g(x)$ approximation function uses a linear combination of a finite number of $g_k(x)$ primitive functions that allow an accurate approximation:

$$g(x) = \sum_{k=1}^{n} g_k(x). \tag{4.2}$$

The particular forms of $g_k(x)$ primitive functions are mainly correlated to the possibility of their facile implementation in CMOS technology. From this perspective, the squaring and multiplying/dividing functions can be realized using simplest CMOS computational structures and, in consequence, primitive functions that can be implemented using these circuits are preferred from economical reasons.

A primitive function is characterized by the number of coefficients that contribute to its definition, these coefficients being used for obtaining the superior-order approximation of the required function.

Taking into account the previous considerations, a possible list of ten primitive functions, associated with their superior-order Taylor series expansions, can be developed as follows.

4.2.1 First Primitive Function

The first primitive function has the following expression:

$$g_1(x) = \frac{x}{1+x} \tag{4.3}$$

and its Taylor series expansion can be expressed as

$$g_1(x) \cong x - x^2 + x^3 - x^4 + x^5 - \ldots . \tag{4.4}$$

4.2.2 Second Primitive Function

The expression of the second primitive function is

$$g_2(x) = \frac{x}{1-x}, \tag{4.5}$$

while the expression of its Taylor series expansion can be written as

$$g_2(x) \cong x + x^2 + x^3 + x^4 + x^5 + \ldots . \tag{4.6}$$

4.2.3 Third Primitive Function

The third primitive function has the following expression:

$$g_3(x) = \frac{1}{1-x} \tag{4.7}$$

and its Taylor series expansion can be expressed as

$$g_3(x) \cong 1 + x + x^2 + x^3 + x^4 + x^5 + \ldots . \tag{4.8}$$

4.2.4 Fourth Primitive Function

The expression of the fourth primitive function is

$$g_4(x) = \frac{1}{1+x}, \tag{4.9}$$

while the expression of its Taylor series expansion can be written as

$$g_4(x) \cong 1 - x + x^2 - x^3 + x^4 - x^5 + \ldots. \tag{4.10}$$

4.2.5 Fifth Primitive Function

The fifth primitive function has the following expression:

$$g_5(x) = \frac{x - 1}{x + 1} \tag{4.11}$$

and its Taylor series expansion can be expressed as

$$g_5(x) - 1 + 2x - 2x^2 + 2x^3 - 2x^4 + 2x^5 - \ldots. \tag{4.12}$$

4.2.6 Sixth Primitive Function

The expression of the sixth primitive function is

$$g_6(x) = \frac{x + 1}{x - 1}, \tag{4.13}$$

while the expression of its Taylor series expansion can be written as

$$g_6(x) \cong -1 - 2x - 2x^2 - 2x^3 - 2x^4 - 2x^5 - \ldots. \tag{4.14}$$

4.2.7 Seventh Primitive Function

The seventh primitive function has the following expression:

$$g_7(x) = \frac{1 + ax}{1 + bx} \tag{4.15}$$

and its Taylor series expansion can be expressed as

$$g_7(x) \cong 1 + (a - b)x + b(b - a)x^2 + b^2(a - b)x^3 + b^3(b - a)x^4 + b^4(a - b)x^5 + \ldots \tag{4.16}$$

4.2.8 Eighth Primitive Function

The expression of the eighth primitive function is

$$g_8(x) = \frac{ax}{1 + bx}, \tag{4.17}$$

while the expression of its Taylor series expansion can be written as

$$g_8(x) \cong ax - abx^2 + ab^2x^3 - ab^3x^4 + ab^4x^5 - \ldots. \tag{4.18}$$

4.2.9 Nineth Primitive Function

The nineth primitive function has the following expression:

$$g_9(x) = \frac{1}{1 + ax^2} \tag{4.19}$$

and its Taylor series expansion can be expressed as

$$f_9(x) \cong 1 - ax^2 + a^2x^4 - a^3x^6 + \ldots. \tag{4.20}$$

4.2.10 Tenth Primitive Function

The expression of the tenth primitive function is

$$g_{10}(x) = \frac{1}{1 + ax}, \tag{4.21}$$

while the expression of its Taylor series expansion can be written as

$$g_{10}(x) \cong 1 - ax + a^2x^2 - a^3x^3 + a^4x^4 - a^6x^6 + \ldots. \tag{4.22}$$

4.3 First Approximation Function

4.3.1 Approximation Function

The first approximation function can be developed using a single $g_{10}(x)$ primitive function:

$$g(x) = bg_{10}(x) + cx + d, \tag{4.23}$$

a, b, c, and d being parameters determined from the condition of third-order approximation of $f(x)$ function using $g(x)$ function:

$$a = -\frac{q}{p}, \tag{4.24}$$

$$b = \frac{p^3}{q^2}, \tag{4.25}$$

$$c = n - \frac{p^2}{q} \tag{4.26}$$

and

$$d = m - \frac{p^3}{q^2}, \tag{4.27}$$

resulting the following equivalent general expression of $g(x)$ approximation function:

$$g(x) = \frac{p^3}{q^2}\frac{1}{1 - \frac{q}{p}x} + \left(n - \frac{p^2}{q}\right)x + m - \frac{p^3}{q^2}. \tag{4.28}$$

The approximation error is mainly caused by the fourth-order term, neglected in the previous approximation:

$$\varepsilon_{f(x)}^{g(x)} \cong \frac{q^2 - rp}{pf(x)}x^4. \tag{4.29}$$

4.3.2 CMOS Implementation of the Function Synthesizer

Block diagram of the function synthesizer

The block diagram of the first function synthesizer, based on the first approximation function (4.23), is presented in the Fig. 4.1. The "SQ" circuit is a current-mode squaring structure, having implementation and description of its operation presented in the following paragraph.

The expression of I_G current is

$$I_G = 2I_F + \frac{I_E^2}{8I_F}, \tag{4.30}$$

where the input currents of this block are

$$I_E = I_O \tag{4.31}$$

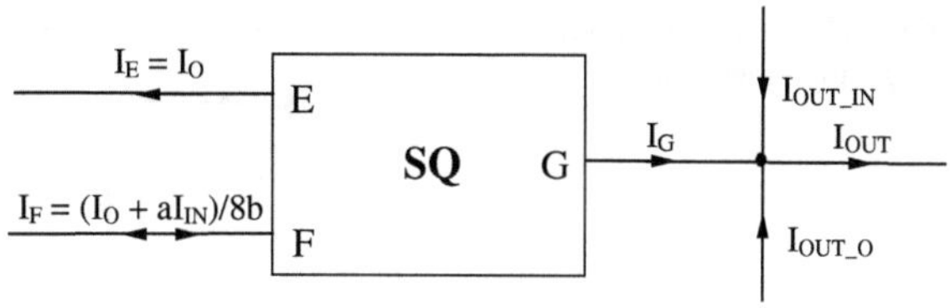

Fig. 4.1 Block diagram of the first function synthesizer

and

$$I_F = \frac{I_O + aI_{IN}}{8b}. \tag{4.32}$$

The additional output currents can be expressed as follows:

$$I_{OUT\_IN} = \left(c - \frac{a}{4b}\right) I_{IN} \tag{4.33}$$

and

$$I_{OUT\_O} = \left(d - \frac{1}{4b}\right) I_O. \tag{4.34}$$

The output current of the circuit having the block diagram presented in Fig. 4.1 will have the following expression:

$$I_{OUT} = I_G + I_{OUT\_IN} + I_{OUT\_O}. \tag{4.35}$$

Using the notation $x = I_{IN}/I_O$ and relation (4.23), it results that I_{OUT} current represents the third-order approximation of $f(x)$ function:

$$I_{OUT} = I_O\left(\frac{b}{1+ax} + cx + d\right) = I_O g(x) \cong I_O f(x). \tag{4.36}$$

Implementation of the squaring block (SQ)
The core for implementing the function synthesizer circuit is represented by a current-mode squaring circuit, having the realization presented in Fig. 4.2, while its symbolical representation is shown in Fig. 4.3.

The input and the output currents can be expressed as $I_E = I - I'$ and $I_G = I + I'$. The translinear loop containing M1 ÷ M4 transistors has the following characteristic equation:

$$V_{SG1} + V_{SG4} = V_{SG2} + V_{SG4}. \tag{4.37}$$

Imposing a biasing in saturation of MOS transistors, it results as follows:

$$2\sqrt{I_F} = \sqrt{\frac{I_G + I_E}{2}} + \sqrt{\frac{I_G - I_E}{2}}. \tag{4.38}$$

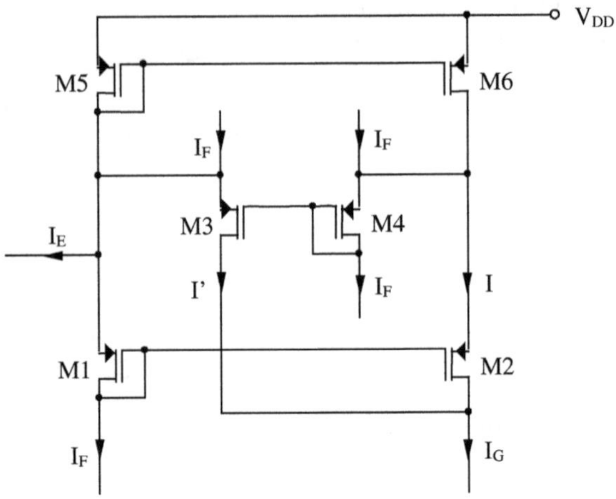

Fig. 4.2 Squaring circuit (SQ) implementation

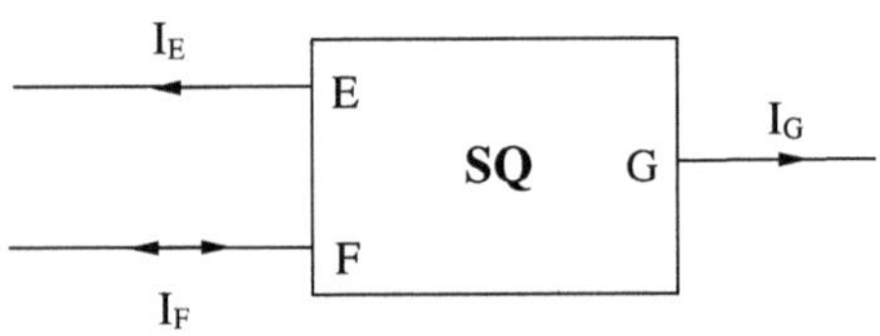

Fig. 4.3 Symbolical representation of the squaring circuit (SQ)

The circuit presented in Fig. 4.2 generates an output current that contains a term proportional with the square of the input current and independent on technological parameters and on temperature variations:

$$I_G = 2I_F + \frac{I_E^2}{8I_F}. \tag{4.39}$$

Implementation of the function synthesizer
The implementation of the function synthesizer circuit, based on the block diagram presented in Fig. 4.1 and on the realization of the squaring circuit from Fig. 4.2, is presented in Fig. 4.4.

4.4 Second Approximation Function

4.4.1 Approximation Function

The second approximation function uses $g_8(x)$ primitive function and linear correction terms in order to generate a multitude of continuous functions:

$$g(x) = g_8(x) + cx + d. \tag{4.40}$$

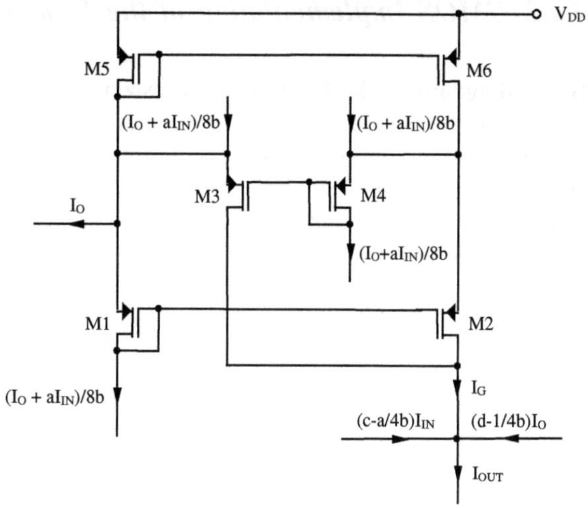

Fig. 4.4 First function synthesizer circuit implementation

The expressions of a, b, c, and d parameters can be determined from the condition of third-order approximation of $f(x)$ function using $g(x)$ function:

$$a = \frac{p^2}{q}, \tag{4.41}$$

$$b = -\frac{q}{p}, \tag{4.42}$$

$$c = n - \frac{p^2}{q} \tag{4.43}$$

and

$$d = m. \tag{4.44}$$

It results the following equivalent general expression of $g(x)$ approximation function:

$$g(x) = \frac{p^2}{q} \frac{x}{1 - \frac{q}{p}x} + \left(n - \frac{p^2}{q} \right) x + m. \tag{4.45}$$

The fourth-order term, neglected in the previous approximation, mainly imposes the approximation error:

$$\varepsilon_{f(x)}^{g(x)} \cong \frac{q^2 + rp}{p f(x)} x^4. \tag{4.46}$$

4.4.2 CMOS Implementation of the Function Synthesizer

Block diagram of the function synthesizer
Using on the second approximation function (4.40), the block diagram of the second function synthesizer is presented in Fig. 4.5. The "MD" circuit is a current-mode multiplier/divider structure, having implementation and description of its operation presented in the following paragraph.

The expression of I_D current is

$$I_D = I_B \frac{I_A}{I_C}, \tag{4.47}$$

where the input currents of this block are

$$I_A = aI_{\text{IN}}, \tag{4.48}$$

$$I_B = I_O \tag{4.49}$$

and

$$I_C = I_O + bI_{\text{IN}}. \tag{4.50}$$

The additional output currents can be expressed as follows:

$$I_{\text{OUT\_IN}} = cI_{\text{IN}} \tag{4.51}$$

and

$$I_{\text{OUT\_O}} = dI_O. \tag{4.52}$$

The output current of the circuit having the block diagram presented in Fig. 4.5 will have the following expression:

$$\begin{aligned} I_{\text{OUT}} &= I_D + I_{\text{OUT\_IN}} + I_{\text{OUT\_O}} \\ &= I_O\left(\frac{ax}{1+bx} + cx + d\right) = I_O g(x) \cong I_O f(x). \end{aligned} \tag{4.53}$$

So, I_{OUT} current will approximate the $f(x)$ function.

Implementation of the squaring block (SQ)
The core for implementing the multiplier/divider circuit is represented by a current-mode squaring circuit, having the realization presented in Fig. 4.6.

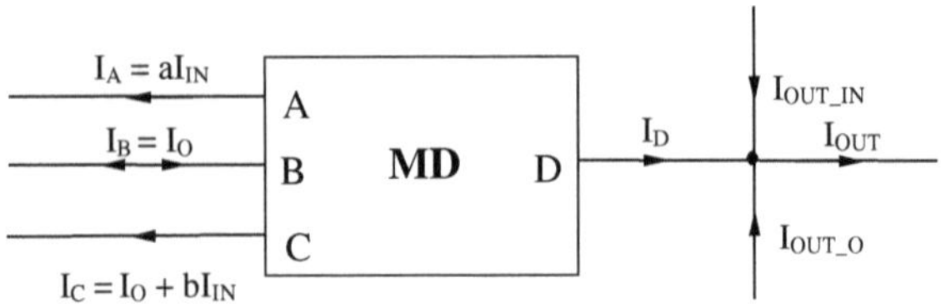

Fig. 4.5 Block diagram of the second function synthesizer

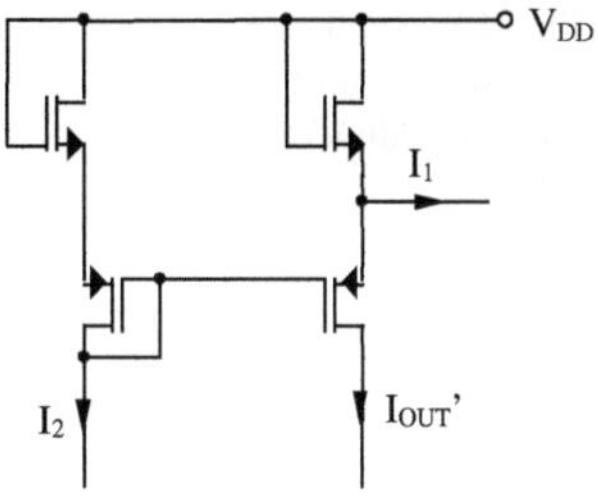

Fig. 4.6 Squaring circuit (SQ) implementation

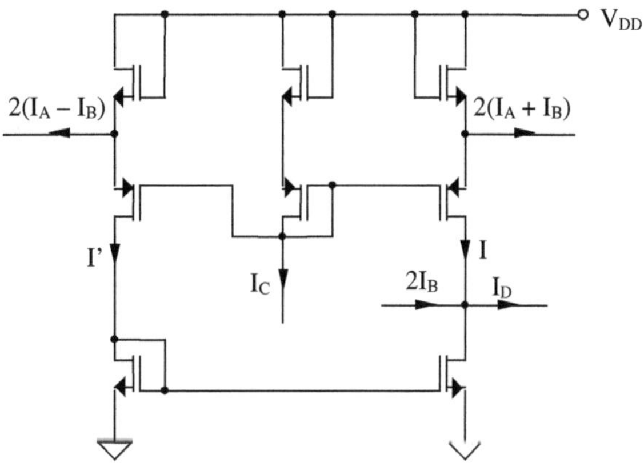

Fig. 4.7 Multiplier/divider circuit (MD) implementation

Noting with $V_{GS}(I)$ the absolute value of the gate–source voltage of a MOS transistor biased at a drain current equal with I, the equation of the translinear loop can be expressed as follows:

$$2V_{GS}(I_2) = V_{GS}(I_{OUT'}) + V_{GS}(I_{OUT'} + I_1), \tag{4.54}$$

resulting

$$2\sqrt{I_2} = \sqrt{I_{OUT'}} + \sqrt{I_{OUT'} + I_1}. \tag{4.55}$$

So

$$I_{OUT'} = I_2 - \frac{I_1}{2} + \frac{I_1^2}{16I_2}. \tag{4.56}$$

Implementation of the multiplier/divider structure (MD)
The multiplier/divider circuit, required by the implementation of the analog function synthesizer circuit, has the realization shown in Fig. 4.7, while its symbolical representation is presented in Fig. 4.8.

The output current of the multiplier circuit has the following expression:

$$I_D = I - I' + 2I_B, \tag{4.57}$$

Fig. 4.8 Symbolical representation of the multiplier/divider circuit (MD)

where

$$I = I_C - (I_A + I_B) + \frac{4(I_A + I_B)^2}{16I_C} \tag{4.58}$$

and

$$I' = I_C - (I_A - I_B) + \frac{4(I_A - I_B)^2}{16I_C}, \tag{4.59}$$

So

$$I_D = \frac{I_A I_B}{I_C}. \tag{4.60}$$

Implementation of the function synthesizer

The implementation of the function synthesizer circuit, based on the block diagram presented in Fig. 4.5 and on the multiplier/divider circuit from Fig. 4.7, is presented in Fig. 4.9.

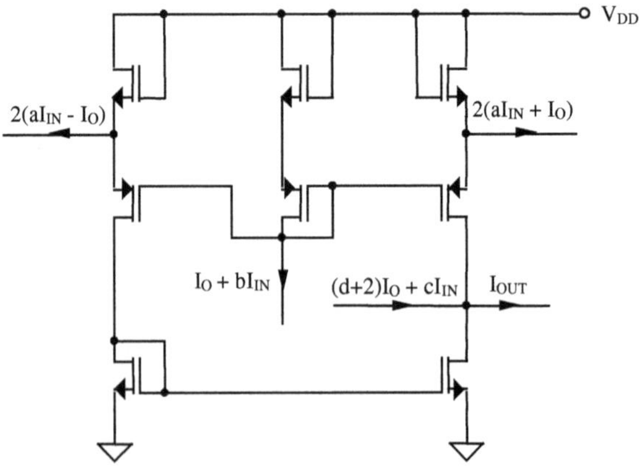

Fig. 4.9 Second function synthesizer circuit implementation

4.5 Third Approximation Function

4.5.1 Approximation Function

Another possible form of the third approximation function can be developed using a single primitive function, $g_7(x)$ and two linear terms:

$$g(x) = g_7(x) + cx + d, \tag{4.61}$$

a, b, c, and d being parameters determined from the condition of third-order approximation of $f(x)$ function using $g(x)$ function:

$$a = \frac{p^2}{q} - \frac{q}{p}, \tag{4.62}$$

$$b = -\frac{q}{p}, \tag{4.63}$$

$$c = n - \frac{p^2}{q} \tag{4.64}$$

and

$$d = m - 1. \tag{4.65}$$

So, it results the following equivalent general expression of $g(x)$ approximation function:

$$g(x) = \frac{1 + \left(\frac{p^2}{q} - \frac{q}{p}\right)x}{1 - \frac{q}{p}x} + \left(n - \frac{p^2}{q}\right)x + (m - 1). \tag{4.66}$$

The most important part of the approximation error is produced by the fourth-order term, neglected in the previous approximation:

$$\varepsilon_{f(x)}^{g(x)} \cong \frac{r - p}{f(x)} x^4. \tag{4.67}$$

4.5.2 CMOS Implementation of the Function Synthesizer

Block diagram of the function synthesizer

The block diagram of the third function synthesizer, based on the third approximation function (4.61), is presented in Fig. 4.10. The "MD" circuit is a current-mode multiplier/divider structure, having implementation and description of its operation presented in the following paragraph.

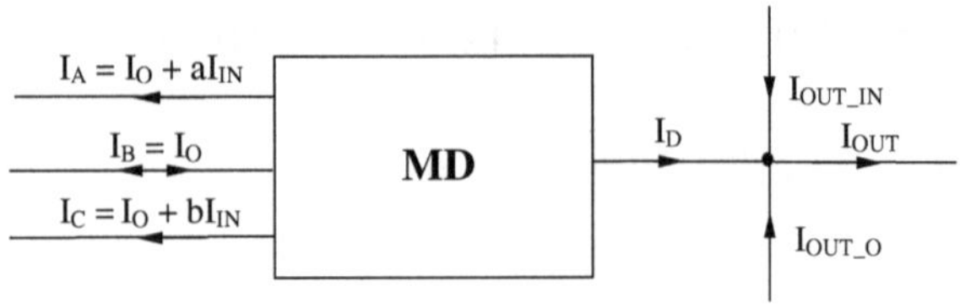

Fig. 4.10 Block diagram of the third function synthesizer

The expression of I_D current is

$$I_D = I_B \frac{I_A}{I_C}. \tag{4.68}$$

The expressions of the input currents of this block are

$$I_A = I_O + aI_{\text{IN}}, \tag{4.69}$$

$$I_B = I_O \tag{4.70}$$

and

$$I_C = I_O + bI_{\text{IN}}. \tag{4.71}$$

The additional output currents can be expressed as follows:

$$I_{\text{OUT\_IN}} = cI_{\text{IN}} \tag{4.72}$$

and

$$I_{\text{OUT\_O}} = dI_O. \tag{4.73}$$

The output current of the circuit having the block diagram presented in Fig. 4.10 will have the following expression:

$$I_{\text{OUT}} = I_D + I_{\text{OUT\_IN}} + I_{\text{OUT\_O}}. \tag{4.74}$$

Considering relation (4.61), it results that I_{OUT} current represents the third-order approximation of $f(x)$ function:

$$I_{\text{OUT}} = I_O\left(\frac{1 + ax}{1 + bx} + cx + d\right) = I_O g(x) \cong I_O f(x). \tag{4.75}$$

Implementation of the squaring block (SQ)
In order to implement the multiplier/divider circuit required by the implementation of function synthesizer circuit, the following current-mode squaring circuit is developed (Fig. 4.11 [1]).

The equation of the translinear loop from Fig. 4.11 can be expressed as follows:

$$2V_{\text{GS}}(I_2) = V_{\text{GS}}(I) + V_{\text{GS}}(I + I_1), \tag{4.76}$$

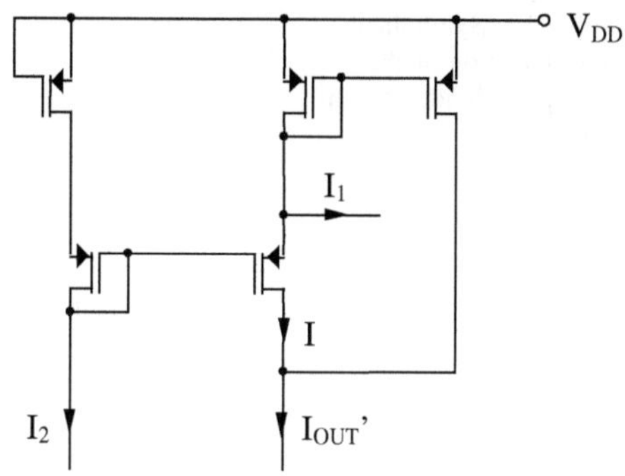

Fig. 4.11 Squaring circuit (SQ) implementation [1]

resulting

$$2\sqrt{I_2} = \sqrt{I} + \sqrt{I + I_1}. \tag{4.77}$$

So

$$I = I_2 - \frac{I_1}{2} + \frac{I_1^2}{16I_2}. \tag{4.78}$$

The expression of $I_{OUT'}$ current will be

$$I_{OUT'} = 2I + I_1 = 2I_2 + \frac{I_1^2}{8I_2}. \tag{4.79}$$

Implementation of the multiplier/divider structure (MD)
The core for implementing the function synthesizer circuit is represented by a multiplier/divider circuit, having the realization presented in Fig. 4.12 [1], while its symbolical representation is shown in Fig. 4.13.

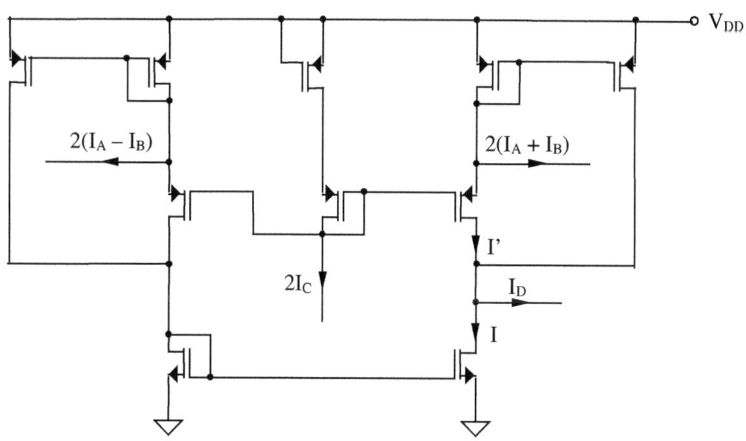

Fig. 4.12 Multiplier/divider circuit (MD) implementation [1]

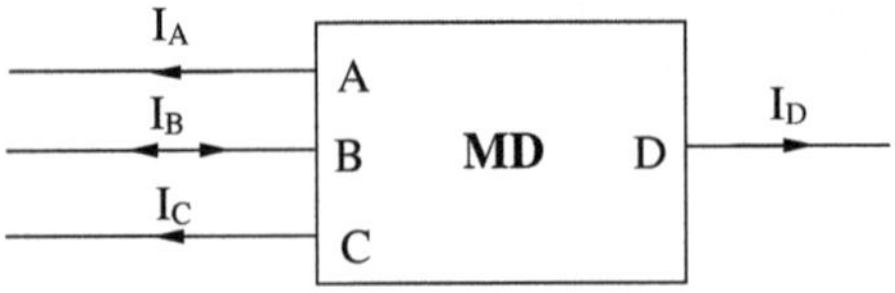

Fig. 4.13 Symbolical representation of the multiplier/divider circuit (MD)

The expressions of I and I' currents are

$$I = 4I_C + \frac{4(I_A - I_B)^2}{16I_C} \tag{4.80}$$

and

$$I' = 4I_C + \frac{4(I_A + I_B)^2}{16I_C}. \tag{4.81}$$

The expression of I_D current will be

$$I_D = I' - I = \frac{I_A I_B}{I_C} \tag{4.82}$$

Implementation of the function synthesizer

The implementation of the function synthesizer circuit presented in Fig. 4.14 [1] uses the block diagram from Fig. 4.10 and the realization of the multiplier/divider circuit shown in Fig. 4.12.

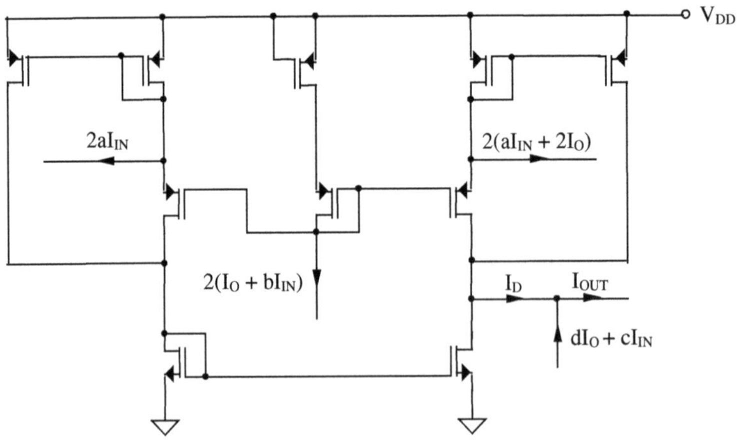

Fig. 4.14 Third function synthesizer circuit implementation [1]

4.6 Fourth Approximation Function

4.6.1 Approximation Function

The fourth approximation function can be developed using a single primitive function, $g_3(x)$, a squaring and two linear terms:

$$g(x) = ag_3(x) + bx^2 + cx + d. \tag{4.83}$$

The *a*, *b*, *c*, and *d* parameters are determined from the condition of third-order approximation of $f(x)$ function using $g(x)$ function, and they have the following expressions:

$$a = q, \tag{4.84}$$

$$b = p - q, \tag{4.85}$$

$$c = n - q \tag{4.86}$$

and

$$d = m - q. \tag{4.87}$$

resulting the following equivalent general expression of $g(x)$ approximation function:

$$g(x) = \frac{q}{1-x} + (p-q)x^2 + (n-q)x + (m-q). \tag{4.88}$$

The approximation error has the following approximate expression:

$$\varepsilon_{f(x)}^{g(x)} \cong \frac{q-r}{f(x)} x^4. \tag{4.89}$$

4.6.2 CMOS Implementation of the Function Synthesizer

Block diagram of the function synthesizer

The fourth function synthesizer using the approximation function expressed by (4.83) has the block diagram presented in Fig. 4.15. The "SQ" circuit is a current-mode squaring structure, having implementation and description of its operation presented in the following paragraph.

The expressions of I_{Ga} and I_{Gb} currents are

$$I_{Ga} = \frac{I_{Fa}}{4} - \frac{I_{Ea}}{2} + \frac{I_{Ea}^2}{4I_{Fa}} \tag{4.90}$$

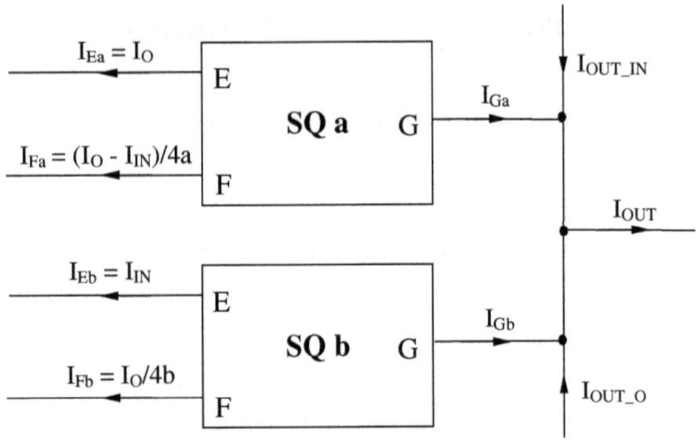

Fig. 4.15 Block diagram of the fourth function synthesizer

and

$$I_{\mathrm{Gb}} = \frac{I_{\mathrm{Fb}}}{4} - \frac{I_{\mathrm{Eb}}}{2} + \frac{I_{\mathrm{Eb}}^2}{4I_{\mathrm{Fb}}}, \tag{4.91}$$

where the input currents of these blocks are

$$I_{\mathrm{Ea}} = I_O, \tag{4.92}$$

$$I_{\mathrm{Fa}} = \frac{I_O - I_{\mathrm{IN}}}{4a}, \tag{4.93}$$

$$I_{\mathrm{Eb}} = I_{\mathrm{IN}} \tag{4.94}$$

and

$$I_{\mathrm{Fb}} = \frac{I_O}{4b}. \tag{4.95}$$

The additional output currents can be expressed as follows:

$$I_{\mathrm{OUT\_IN}} = \left(c + \frac{1}{16a} + \frac{1}{2}\right) I_{\mathrm{IN}} \tag{4.96}$$

and

$$I_{\mathrm{OUT\_O}} = \left(d - \frac{1}{16a} - \frac{1}{16b} + \frac{1}{2}\right) I_O. \tag{4.97}$$

The output current of the circuit having the block diagram presented in Fig. 4.15 will have the following expression:

$$I_{\mathrm{OUT}} = I_{\mathrm{Ga}} + I_{\mathrm{Gb}} + I_{\mathrm{OUT\_IN}} + I_{\mathrm{OUT\_O}}. \tag{4.98}$$

Using relation (4.83), it results that I_{OUT} current represents the third-order approximation of $f(x)$ function:

$$I_{\mathrm{OUT}} = I_O\left(\frac{a}{1-x} + bx^2 + cx + d\right) = I_O g(x) \cong I_O f(x). \tag{4.99}$$

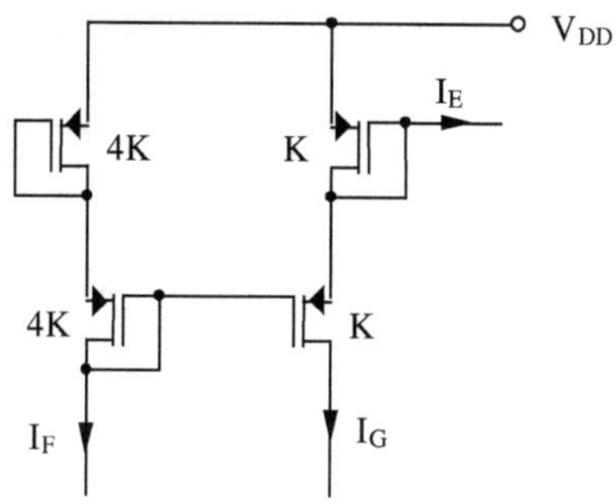

Fig. 4.16 Squaring circuit (SQ) implementation

Implementation of the squaring block (SQ)
The current-mode squaring circuit has the realization presented in Fig. 4.16, and its symbolical representation is shown in Fig. 4.17.

The translinear loop that establishes the operation of the functional squaring circuit proposed in Fig. 4.16 can be expressed as follows:

$$2V_{GS}(I_F, 4K) = V_{GS}(I_G, K) + V_{GS}(I_G + I_E, K), \qquad (4.100)$$

resulting

$$I_G = \frac{I_F}{4} - \frac{I_E}{2} + \frac{I_E^2}{4I_F}. \qquad (4.101)$$

Implementation of the function synthesizer
The implementation of the function synthesizer circuit, based on the block diagram presented in Fig. 4.15 and on the realization of the squaring circuit from Fig. 4.16, is presented in Fig. 4.18.

4.7 Fifth Approximation Function

4.7.1 Approximation Function

The fifth approximation function can be expressed as follows:

$$g(x) = ag_4(x) + bx^2 + cx + d, \qquad (4.102)$$

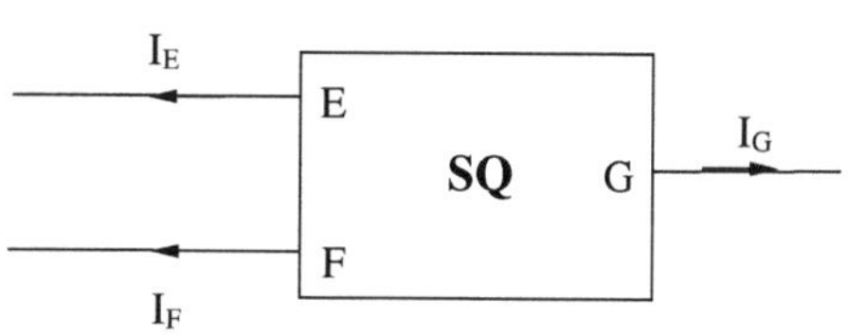

Fig. 4.17 Symbolical representation of the squaring circuit (SQ)

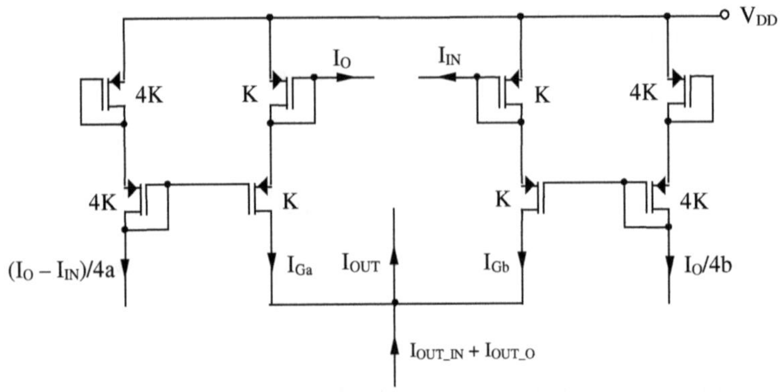

Fig. 4.18 Fourth function synthesizer circuit implementation

a, b, c, and d being parameters determined from the condition of third-order approximation of $f(x)$ function using $g(x)$ function:

$$a = -q, \tag{4.103}$$

$$b = p + q, \tag{4.104}$$

$$c = n - q \tag{4.105}$$

and

$$d = m + q. \tag{4.106}$$

So, the equivalent general expression of $g(x)$ approximation function can be written as

$$g(x) = -\frac{q}{1+x} + (p+q)x^2 + (n-q)x + (m+q). \tag{4.107}$$

The approximation error is approximately

$$\varepsilon_{f(x)}^{g(x)} \cong \frac{q+r}{f(x)}x^4. \tag{4.108}$$

4.7.2 CMOS Implementation of the Function Synthesizer

Block Diagram of the Function Synthesizer

The block diagram of the fifth function synthesizer that uses the fifth approximation function (4.102) is shown in Fig. 4.19. The "SQ" circuit is a current-mode squaring structure, having implementation and description of its operation presented in the following paragraph.

Fig. 4.19 Block diagram of the fifth function synthesizer

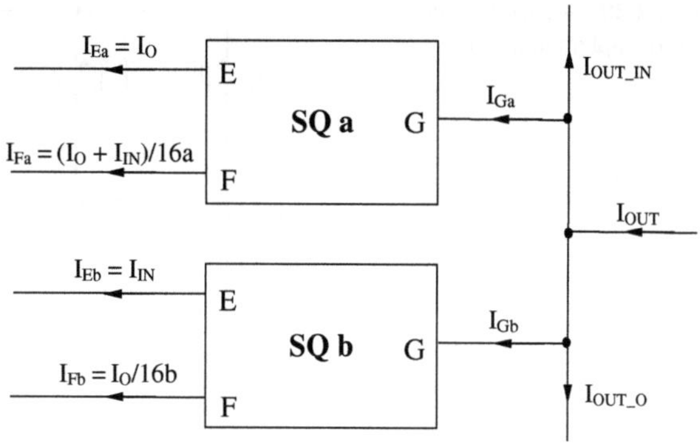

The expressions of I_{Ga} and I_{Gb} currents are

$$I_{Ga} = \frac{I_{Ea}^2}{16 I_{Fa}} \quad (4.109)$$

and

$$I_{Gb} = \frac{I_{Eb}^2}{16 I_{Fb}}, \quad (4.110)$$

where the input currents of these blocks are

$$I_{Ea} = I_O, \quad (4.111)$$

$$I_{Fa} = \frac{I_O + I_{IN}}{16a}, \quad (4.112)$$

$$I_{Eb} = I_{IN} \quad (4.113)$$

and

$$I_{Fb} = \frac{I_O}{16b}. \quad (4.114)$$

The additional output currents have the following expressions:

$$I_{OUT\_IN} = cI_{IN} \quad (4.115)$$

and

$$I_{OUT\_O} = dI_O. \quad (4.116)$$

The output current of the circuit presented in Fig. 4.19 will be

$$I_{OUT} = I_{Ga} + I_{Gb} + I_{OUT\_IN} + I_{OUT\_O}. \quad (4.117)$$

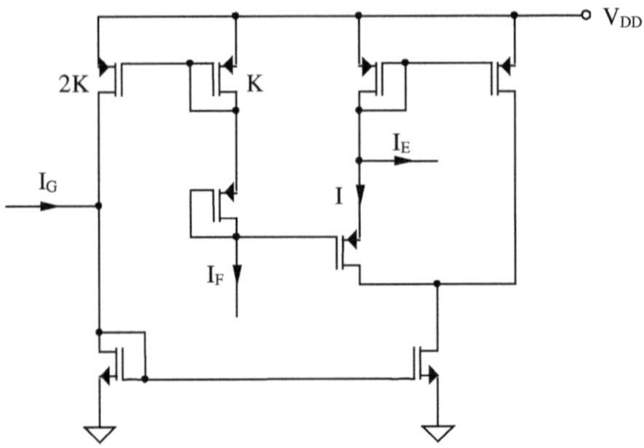

Fig. 4.20 Squaring circuit (SQ) implementation

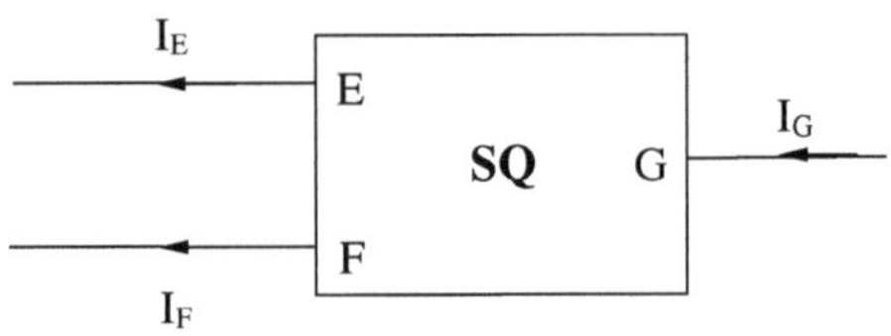

Fig. 4.21 Symbolical representation of the squaring circuit (SQ)

Considering relation (4.102), it can conclude that I_{OUT} current represents the third-order approximation of $f(x)$ function:

$$I_{\text{OUT}} = I_O\left(\frac{a}{1+x} + bx^2 + cx + d\right) = I_O g(x) \cong I_O f(x). \tag{4.118}$$

Implementation of the Squaring block (SQ)

The implementation of the current-mode squaring circuit is presented in Fig. 4.20, while its symbolical representation is shown in Fig. 4.21.

For the previous squaring circuit, the equation of the translinear loop can be expressed as follows:

$$2V_{\text{GS}}(I_F) = V_{\text{GS}}(I) + V_{\text{GS}}(I + I_E), \tag{4.119}$$

resulting

$$2\sqrt{I_F} = \sqrt{I} + \sqrt{I + I_E}. \tag{4.120}$$

So

$$I = I_F - \frac{I_E}{2} + \frac{I_E^2}{16 I_F}. \tag{4.121}$$

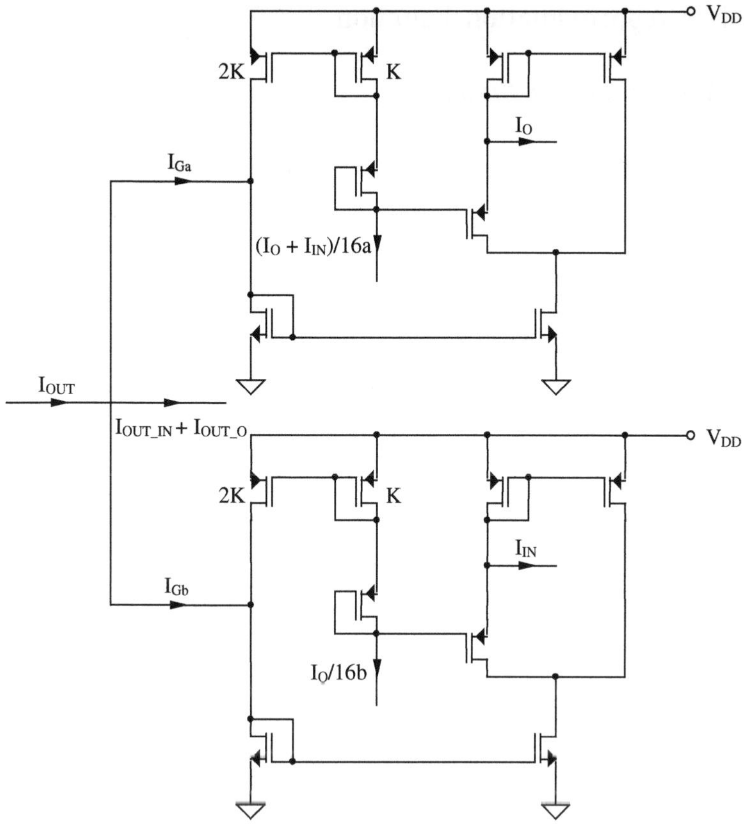

Fig. 4.22 Fifth function synthesizer circuit implementation

The expression of the output current of the squaring circuit will be

$$I_G = 2I + I_E - 2I_F = \frac{I_E^2}{16I_F}. \tag{4.122}$$

Implementation of the function synthesizer

The implementation of the function synthesizer circuit, based on the block diagram presented in Fig. 4.19 and on the realization of the squaring circuit from Fig. 4.20, is presented in Fig. 4.22.

4.8 Sixth Approximation Function

4.8.1 Approximation Function

The sixth approximation function can be developed using two primitive functions, $g_3(x)$ and $g_4(x)$:

$$g(x) = ag_3(x) + bg_4(x) + cx + d. \tag{4.123}$$

The condition of third-order approximation of $f(x)$ function using $g(x)$ function will impose the expressions of a, b, c, and d being parameters:

$$a = \frac{p+q}{2}, \tag{4.124}$$

$$b = \frac{p-q}{2}, \tag{4.125}$$

$$c = n - q \tag{4.126}$$

and

$$d = m - p. \tag{4.127}$$

So, the equivalent general expression of $g(x)$ approximation function will be

$$g(x) = \frac{p+q}{2}\frac{1}{1-x} - \frac{p-q}{2}\frac{1}{1+x} + (n-q)x + (m-p). \tag{4.128}$$

The approximation error is mainly caused by the fourth-order term, neglected in the previous approximation:

$$\varepsilon_{f(x)}^{g(x)} = \frac{r-p}{f(x)}x^4. \tag{4.129}$$

4.8.2 CMOS Implementation of the Function Synthesizer

Block diagram of the function synthesizer

The block diagram of the sixth function synthesizer is presented in Fig. 4.23.

The I_{Ga} and I_{Gb} currents can be expressed as follows:

$$I_{Ga} = I_{Fa} - \frac{I_{Ea}}{2} + \frac{I_{Ea}^2}{16I_{Fa}} \tag{4.130}$$

and

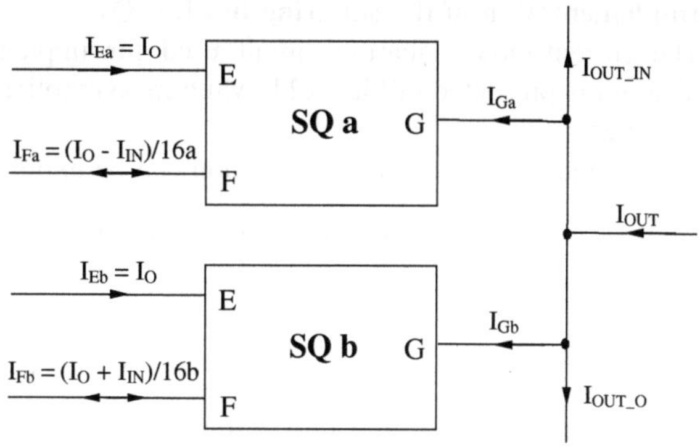

Fig. 4.23 Block diagram of the sixth function synthesizer

$$I_{Gb} = I_{Fb} - \frac{I_{Eb}}{2} + \frac{I_{Eb}^2}{16 I_{Fb}}, \tag{4.131}$$

where the input currents of these blocks are

$$I_{Ea} = I_O, \tag{4.132}$$

$$I_{Fa} = \frac{I_O - I_{IN}}{16a}, \tag{4.133}$$

$$I_{Eb} = I_O \tag{4.134}$$

and

$$I_{Fb} = \frac{I_O + I_{IN}}{16b}. \tag{4.135}$$

The I_{OUT_IN} and I_{OUT_O} currents have the following expressions:

$$I_{OUT\_IN} = \left(c + \frac{1}{16a} - \frac{1}{16b}\right) I_{IN} \tag{4.136}$$

and

$$I_{OUT\_O} = \left(d - \frac{1}{16a} - \frac{1}{16b} + 1\right) I_O. \tag{4.137}$$

The expression of the output current for the circuit from Fig. 4.23 is

$$I_{OUT} = I_{Ga} + I_{Gb} + I_{OUT\_IN} + I_{OUT\_O}. \tag{4.138}$$

Taking into account relation (4.123), it results that I_{OUT} current represents the third-order approximation of $f(x)$ function:

$$I_{OUT} = I_O\left(\frac{a}{1-x} + \frac{b}{1+x} + cx + d\right) = I_O g(x) \cong I_O f(x). \tag{4.139}$$

Implementation of the squaring block (SQ)

The current-mode squaring circuit used for implementing the previous block diagram is presented in Fig. 4.24, while its symbolical representation is shown in Fig. 4.25.

The equation of the translinear loop can be expressed as follows:

$$2V_{GS}(I_F) = V_{GS}(I_G) + V_{GS}(I_G + I_E), \tag{4.140}$$

resulting

$$2\sqrt{I_F} = \sqrt{I_G} + \sqrt{I_G + I_E}. \tag{4.141}$$

The expression of the output current of the squaring circuit will be

$$I_G = I_F - \frac{I_E}{2} + \frac{I_E^2}{16 I_F}. \tag{4.142}$$

Implementation of the function synthesizer

The implementation of the function synthesizer circuit is presented in Fig. 4.26.

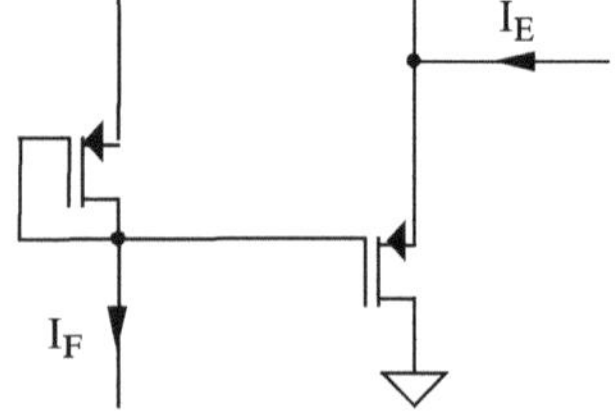

Fig. 4.24 Squaring circuit (SQ) implementation

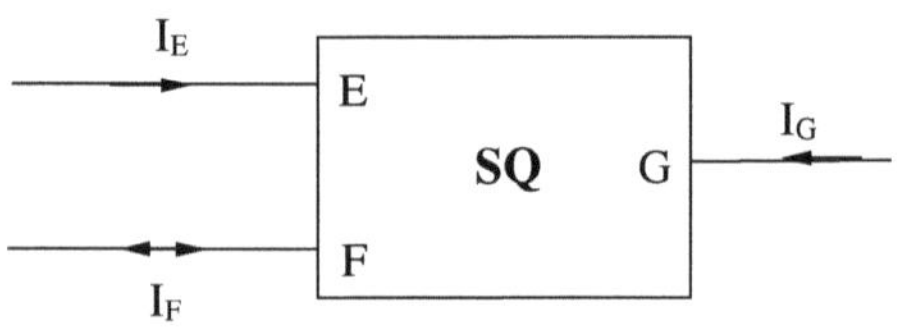

Fig. 4.25 Symbolical representation of the squaring circuit (SQ)

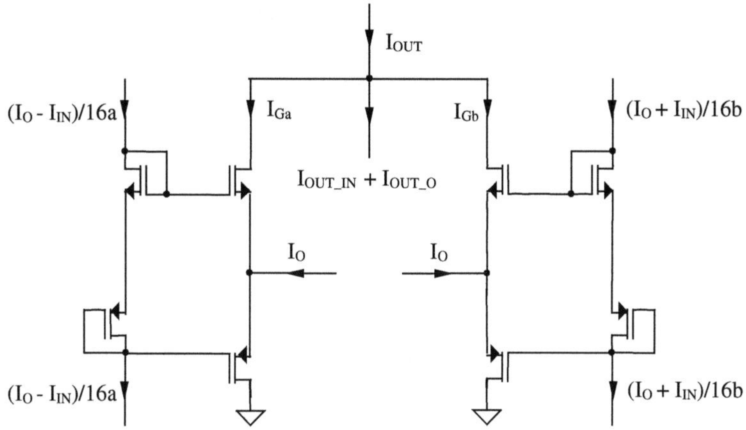

Fig. 4.26 Sixth function synthesizer circuit implementation

Reference

1. Cosmin Popa, "Low-Voltage Improved Accuracy Gaussian Function Generator with Fourth-Order Approximation", Microelectronics Journal, 2012, Volume 43 Issue 8, August, 2012, pp. 515-520.

Chapter 5
Fourth-Order Function Synthesizers

5.1 Introduction

In order to develop analog function synthesizer circuits having an increased accuracy comparing with the third-order computational structures analyzed in the previous chapter, superior-order approximation functions have to be considered. The increase in the order of approximation strongly increases the complexity of the designed circuit, being necessary to realize a trade-off between the circuit complexity and its overall accuracy. From this point of view, a convenient choice that allows to obtain a very good accuracy and a relative large dynamic range of the function synthesizer using a reasonable circuit complexity is based on a fourth-order approximation. This fourth-order approximation is usually enough accurate for generating the most important circuit functions from the perspective of their applications in analog signal processing and of fulfilling the conditions imposed by these applications. Comparing with previously presented computational structures based on third-order approximation functions, the same requirements must be taken into account when design a high-precision fourth-order analog function synthesizer circuit.

The fourth-order match between approximated $f(x)$ function and $g(x)$ approximation function is equivalent with the identity of the first five terms from their expansions: the constant term, the linear term, the second-order term, the third-order term, and the fourth-order term. In order to reduce the complexity of the CMOS circuits that implement the following presented approximation functions, they will be usually developed using only two primitive functions and, in most cases, using two linear additional terms. These terms will not require any supplementary computational circuits, as they can be implemented using linear relations between the currents from the circuit. The fourth-order Taylor series of $f(x)$ function can be considered to have the following general expression:

$$f(x) \cong m + nx + px^2 + qx^3 + rx^4, \tag{5.1}$$

m, n, p, q, and r, being the coefficients of the expansions, determined by the particular expression of $f(x)$ function.

C. R. Popa, *Current-Mode Analog Nonlinear Function Synthesizer Structures*,
DOI: 10.1007/978-3-319-01035-9_5,

5.2 First Function Synthesizer Circuit

5.2.1 Approximation Function

In order to implement an improved accuracy analog function synthesizer circuit, the following presented method is to use the first approximation function, that can be developed using a linear combination of two primitive functions, $g_7(x)$ and $g_1(x)$:

$$g(x) = g_7(x) + cg_1(x) + dx + e, \tag{5.2}$$

a, *b*, *c*, *d*, and *e* being parameters determined from the condition of fourth-order approximation of $f(x)$ function using $g(x)$ function.

The motivation for choosing this particular $g(x)$ approximation function is correlated with the possibilities of its facile implementation in CMOS technology (the multiplier/divider [MD] function has a relatively simple realization). The fourth-order identity between the Taylor series of $f(x)$ and $g(x)$ functions is equivalent with the identity between the first five terms from the Taylor series of the previous functions. It results

$$a = \frac{(p+q)^3}{(r+q)(r+p+2q)} - \frac{r+q}{p+q}, \tag{5.3}$$

$$b = -\frac{r+q}{p+q}, \tag{5.4}$$

$$c = \frac{(p+q)^2}{r+p+2q} - p, \tag{5.5}$$

$$d = n + p - \frac{(p+q)^2}{r+q}, \tag{5.6}$$

$$e = m - 1. \tag{5.7}$$

The approximation error will have the following general expression:

$$\varepsilon_{f(x)}^{g(x)}(x) \cong \left[\frac{q^2 - 2rp - qr}{p+q} - s\right] \frac{x^5}{f(x)}. \tag{5.8}$$

5.2.2 CMOS Implementation of the Function Synthesizer

5.2.2.1 Block Diagram of the Function Synthesizer

The block diagram of the first function synthesizer, based on the first approximation function (5.2), is presented in Fig. 5.1. The "MD" circuits are current-mode MD structures, having implementation and description of their operation presented in the following paragraph.

The expressions of I_{Da} and I_{Db} currents are:

$$I_{Da} = \frac{I_{Ba} I_{Aa}}{I_{Ca}} \tag{5.9}$$

and

$$I_{Db} = \frac{I_{Bb} I_{Ab}}{I_{Cb}}. \tag{5.10}$$

where the input currents of these blocks are:

$$I_{Aa} = I_O + aI_{IN}, \tag{5.11}$$

$$I_{Ba} = I_O, \tag{5.12}$$

$$I_{Ca} = I_O + bI_{IN}, \tag{5.13}$$

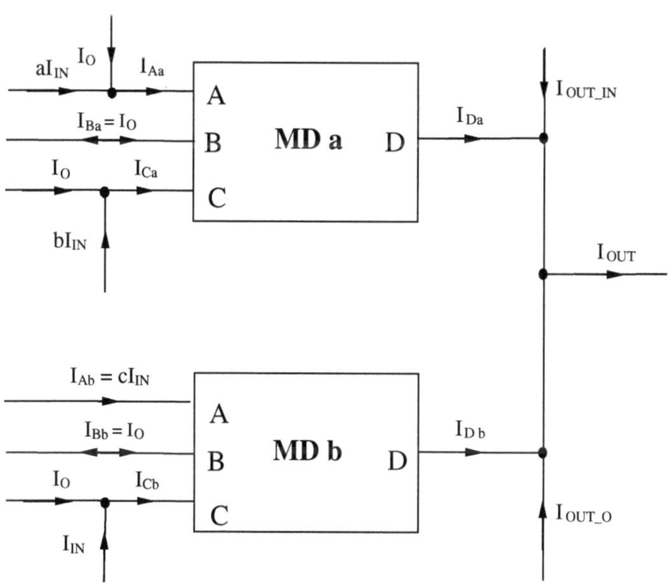

Fig. 5.1 Block diagram of the first function synthesizer

$$I_{Ab} = cI_{\mathrm{IN}}, \tag{5.14}$$

$$I_{Bb} = I_O \tag{5.15}$$

and

$$I_{Cb} = I_O + I_{\mathrm{IN}}. \tag{5.16}$$

The additional output currents can be expressed as follows:

$$I_{\mathrm{OUT\_IN}} = dI_{\mathrm{IN}} \tag{5.17}$$

and

$$I_{\mathrm{OUT\_}O} = eI_O. \tag{5.18}$$

The output current of the circuit having the block diagram presented in Fig. 5.1 will have the following expression:

$$I_{\mathrm{OUT}} = I_{Da} + I_{Db} + I_{\mathrm{OUT\_IN}} + I_{\mathrm{OUT\_}O}. \tag{5.19}$$

Using the notation $x = I_{\mathrm{IN}}/I_O$ and relation (5.2), it results

$$I_{\mathrm{OUT}} = I_O\left(\frac{1+ax}{1+bx} + \frac{cx}{1+x} + dx + e\right). \tag{5.20}$$

So, I_{OUT} current represents the fourth-order approximation of $f(x)$ function:

$$I_{\mathrm{OUT}} = I_O g(x) \cong I_O f(x). \tag{5.21}$$

5.2.2.2 Implementation of the Squaring Block

The functional core for designing the "MD" circuit is represented by a current-mode squaring circuit (SQ), having the implementation presented in Fig. 5.2 [1].

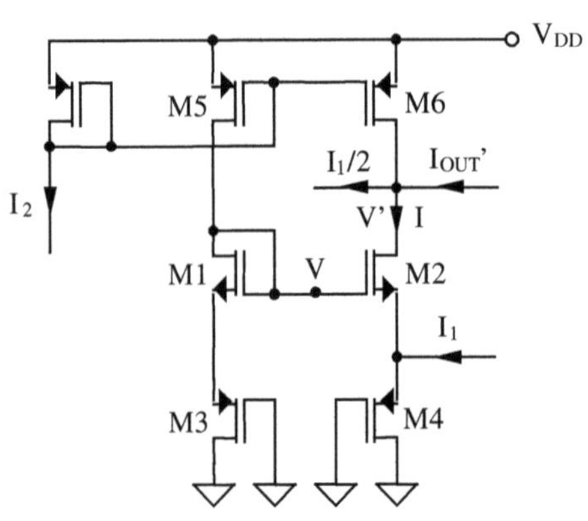

Fig. 5.2 Squaring circuit (*SQ*) implementation [1]

Noting with $V_{GS}(I)$ the absolute value of the gate-source voltage of a MOS transistor biased at a drain current equal with I, the equation of the translinear loop can be expressed as follows:

$$2V_{GS}(I_2) = V_{GS}(I) + V_{GS}(I + I_1), \tag{5.22}$$

resulting

$$2\sqrt{I_2} = \sqrt{I} + \sqrt{I + I_1}. \tag{5.23}$$

So

$$I = I_2 - \frac{I_1}{2} + \frac{I_1^2}{16I_2}. \tag{5.24}$$

The output current can be expressed as follows:

$$I'_{OUT} = I + \frac{I_1}{2} - I_2 = \frac{I_1^2}{16I_2}. \tag{5.25}$$

5.2.2.3 Implementation of the Multiplier/Divider Blocks

The circuit MD, derived from the SQ circuit from Fig. 5.2, is presented in Fig. 5.3 [1]. It is designed using two current-mode SQ circuits (M1–M4 and M3–M6, respectively). The complexity of the "MD" circuit is minimized by reusing M3–M4 transistors for both SQ circuits. The symbolical representation of the MD circuit is shown in Fig. 5.4.

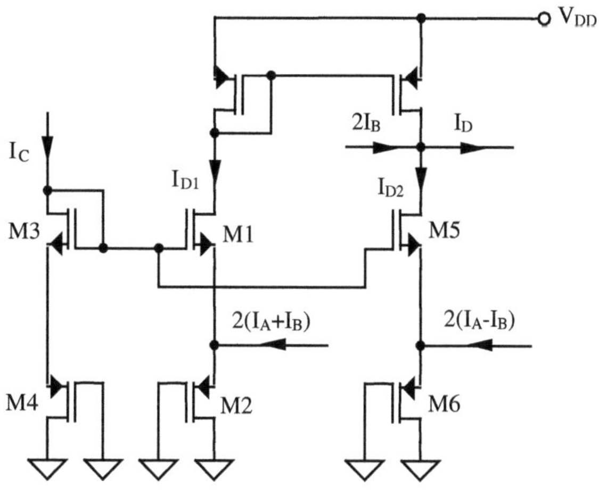

Fig. 5.3 Multiplier/divider circuit (*MD*) implementation [1]

Fig. 5.4 Symbolical representation of the multiplier/divider circuit (*MD*)

The output current has the following expression:

$$I_D = I_{D1} - I_{D2} + 2I_B, \tag{5.26}$$

where two translinear loops similar with the loop analyzed for the "SQ" circuit implement the following expressions of I_{D1} and I_{D2} currents [derived from (5.24)]:

$$I_{D1} = I_C - (I_A + I_B) + \frac{(I_A + I_B)^2}{4I_C} \tag{5.27}$$

and

$$I_{D2} = I_C - (I_A - I_B) + \frac{(I_A - I_B)^2}{4I_C}, \tag{5.28}$$

resulting an output current of the circuit having the following expression:

$$I_D = I_B \frac{I_A}{I_C}. \tag{5.29}$$

The MD circuit presents two important advantages. First, the current-mode operation increases the frequency response of the computational structure. Second, the independence of the output current expressed by (5.29) on technological parameters removes, in a first-order analysis, the temperature and technological-caused errors, the overall accuracy of the proposed circuit being increased in this way.

5.2.2.4 Implementation of the Function Synthesizer

The implementation of the function synthesizer circuit, based on the block diagram presented in Fig. 5.1 and on the realization of the MD circuit from Fig. 5.3 is presented in Fig. 5.5 [1].

5.3 Second Function Synthesizer Circuit

5.3.1 Approximation Function

The second approximation function can be developed using a linear combination of two primitive functions, $g_8(x)$ and $g_2(x)$:

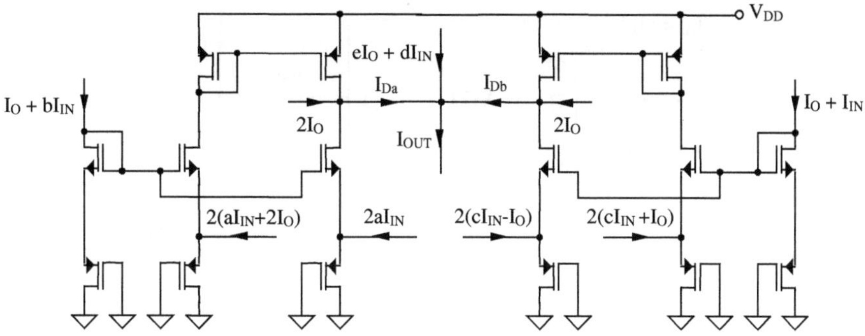

Fig. 5.5 First function synthesizer circuit implementation [1]

$$g(x) = g_8(x) + cg_2(x) + dx + e. \tag{5.30}$$

The condition of fourth-order approximation of $f(x)$ function using $g(x)$ function will impose the expressions of a, b, c, d, and e parameters:

$$a = \frac{(q-p)^3}{(q-r)(2q-r-p)}, \tag{5.31}$$

$$b = \frac{q-r}{q-p}, \tag{5.32}$$

$$c = p + \frac{(q-p)^2}{2q-r-p} \tag{5.33}$$

and

$$d = n - p - \frac{(q-p)^2}{q-r}, \tag{5.34}$$

$$e = m. \tag{5.35}$$

5.3.2 CMOS Implementation of the Function Synthesizer

5.3.2.1 Block Diagram of the Function Synthesizer

The block diagram of the second function synthesizer, using the second approximation function (5.30), is presented in Fig. 5.6.

The "MD" circuits are current-mode MD structures, having implementation and description of their operation presented in the following paragraph. The expressions of I_{Da} and I_{Db} currents are:

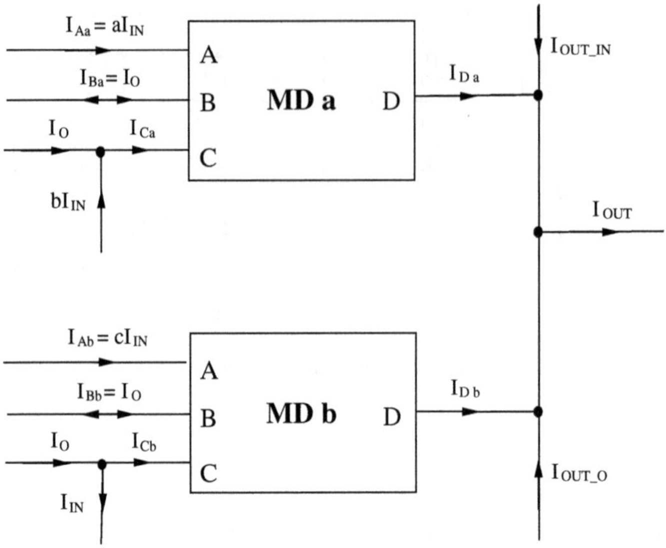

Fig. 5.6 Block diagram of the second function synthesizer

$$I_{Da} = \frac{I_{Ba}I_{Aa}}{I_{Ca}} \tag{5.36}$$

and

$$I_{Db} = \frac{I_{Bb}I_{Ab}}{I_{Cb}}, \tag{5.37}$$

where the input currents of these blocks are:

$$I_{Aa} = aI_{\text{IN}}, \tag{5.38}$$

$$I_{Ba} = I_O, \tag{5.39}$$

$$I_{Ca} = I_O + bI_{\text{IN}}, \tag{5.40}$$

$$I_{Ab} = cI_{\text{IN}}, \tag{5.41}$$

$$I_{Bb} = I_O \tag{5.42}$$

and

$$I_{Cb} = I_O - I_{\text{IN}}. \tag{5.43}$$

The additional output currents can be expressed as follows:

$$I_{\text{OUT\_IN}} = dI_{\text{IN}} \tag{5.44}$$

and

$$I_{\mathrm{OUT\_}O} = eI_O. \tag{5.45}$$

The output current of the circuit having the block diagram presented in Fig. 5.6 will have the following expression:

$$I_{\mathrm{OUT}} = I_{Da} + I_{Db} + dI_{\mathrm{IN}} + eI_O. \tag{5.46}$$

Using relation (5.30), it can conclude that I_{OUT} current represents the fourth-order approximation of $f(x)$ function:

$$I_{\mathrm{OUT}} = I_O\left(\frac{ax}{1+bx} + \frac{cx}{1-x} + dx + e\right) = I_O g(x) \cong I_O f(x). \tag{5.47}$$

5.3.2.2 Implementation of the Multiplier/Divider Block

The implementation of the MD circuit that will be used for obtaining the function generated by the function synthesizer circuit is presented in Fig. 5.7, while its symbolical representation is shown in Fig. 5.8.

The expressions of I and I' currents are:

$$I - I_B - (I_A + I_B) + \frac{(I_A + I_B)^2}{4I_C} \tag{5.48}$$

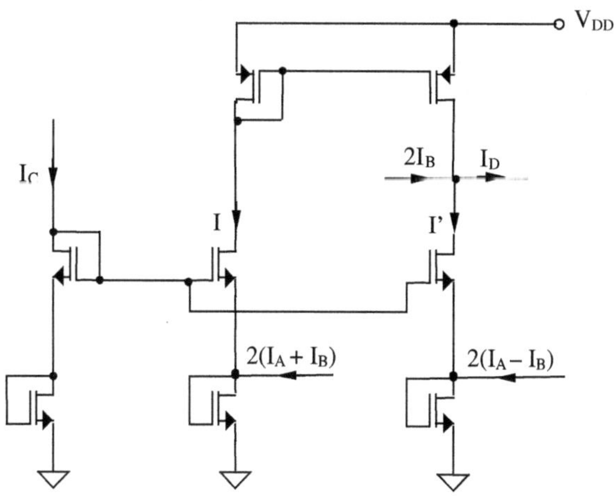

Fig. 5.7 Multiplier/divider circuit (*MD*) implementation

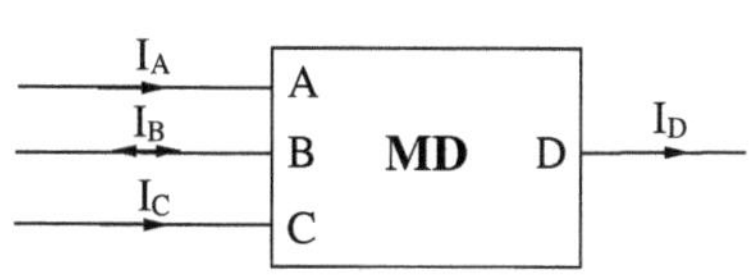

Fig. 5.8 Symbolical representation of the multiplier/divider circuit (*MD*)

and

$$I' = I_B - (I_A - I_B) + \frac{(I_A - I_B)^2}{4I_C}. \tag{5.49}$$

It results

$$I_D = I - I' + 2I_B = \frac{I_A I_B}{I_C}. \tag{5.50}$$

5.3.2.3 Implementation of the Function Synthesizer

The function synthesizer circuit, using the block diagram presented in Fig. 5.6 and the MD circuit implementation from Fig. 5.7, is presented in Fig. 5.9.

5.4 Third Function Synthesizer Circuit

5.4.1 Approximation Function

The third approximation function represents a linear combination of two primitive functions, $g_7(x)$ and $g_2(x)$:

$$g(x) = g_7(x) + cg_2(x) + dx + e. \tag{5.51}$$

The expressions of a, b, c, d, and e parameters can be determined from the condition of fourth-order approximation of $f(x)$ function using $g(x)$ function:

$$a = \frac{q-r}{q-p} + \frac{(q-p)^3}{(q-r)(2q-r-p)}, \tag{5.52}$$

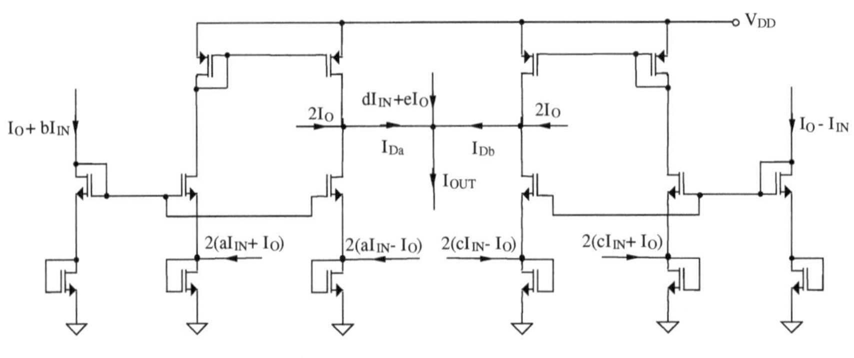

Fig. 5.9 Second function synthesizer circuit implementation

$$b = \frac{q-r}{q-p}, \tag{5.53}$$

$$c = p + \frac{(q-p)^2}{2q-r-p} \tag{5.54}$$

and

$$d = n - p - \frac{(q-p)^2}{q-r}, \tag{5.55}$$

$$e = m - 1. \tag{5.56}$$

5.4.2 CMOS Implementation of the Function Synthesizer

5.4.2.1 Block Diagram of the Function Synthesizer

The block diagram of the fourth function synthesizer, based on the third approximation function (5.51), is presented in Fig. 5.10.

The realization and operation description for "MD" circuits (current-mode MD structures) are presented in the following paragraph. The expressions of I_{Da} and I_{Db} currents are

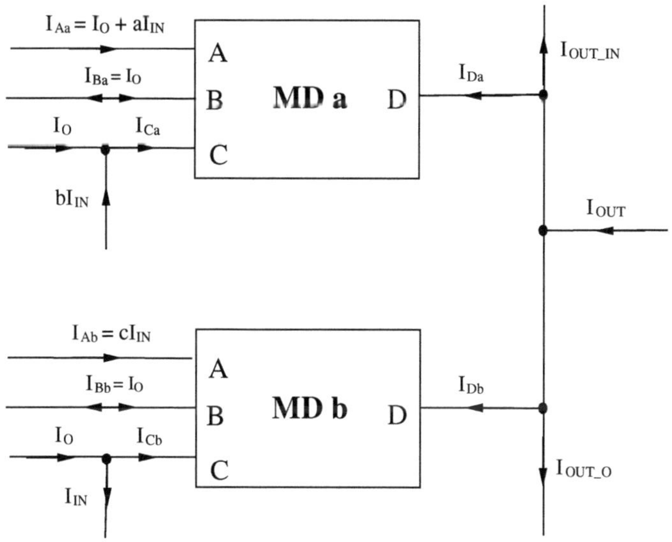

Fig. 5.10 Block diagram of the third function synthesizer

$$I_{Da} = \frac{I_{Ba} I_{Aa}}{I_{Ca}} \tag{5.57}$$

and

$$I_{Db} = \frac{I_{Bb} I_{Ab}}{I_{Cb}}, \tag{5.58}$$

where the input currents of these blocks are

$$I_{Aa} = I_O + aI_{IN}, \tag{5.59}$$

$$I_{Ba} = I_O, \tag{5.60}$$

$$I_{Ca} = I_O + bI_{IN}, \tag{5.61}$$

$$I_{Ab} = cI_{IN}, \tag{5.62}$$

$$I_{Bb} = I_O \tag{5.63}$$

and

$$I_{Cb} = I_O - I_{IN}. \tag{5.64}$$

The expression of the additional output currents are:

$$I_{OUT\_IN} = dI_{IN} \tag{5.65}$$

and

$$I_{OUT\_O} = eI_O. \tag{5.66}$$

The output current of the circuit having the block diagram presented in Fig. 5.10 will have the following expression:

$$I_{OUT} = I_{Da} + I_{Db} + dI_{IN} + eI_O. \tag{5.67}$$

Using relation (5.51), it results that I_{OUT} current represents the fourth-order approximation of $f(x)$ function:

$$I_{OUT} = I_O\left(\frac{1 + ax}{1 + bx} + \frac{cx}{1 - x} + dx + e\right) = I_O g(x) \cong I_O f(x). \tag{5.68}$$

5.4.2.2 Implementation of the Squaring Block

The current-mode SQ circuit realization is presented in Fig. 5.11.

The translinear loop that establishes the operation of the functional SQ circuit from Fig. 5.11 can be expressed as follows:

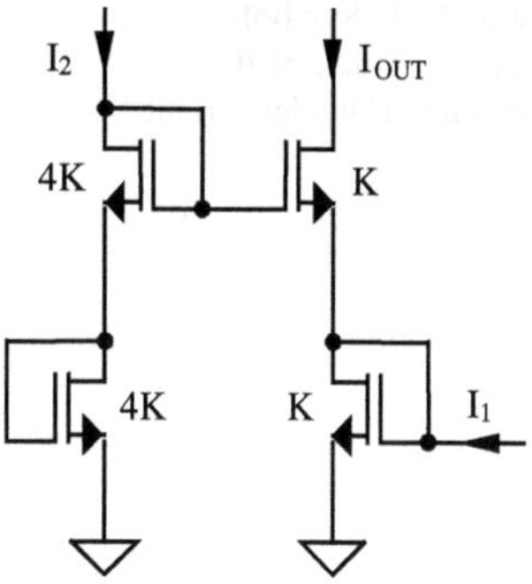

Fig. 5.11 Squaring circuit (*SQ*) implementation

$$2V_{GS}(I_2, 4K) = V_{GS}(I_{\text{OUT}}, K) + V_{GS}(I_{\text{OUT}} + I_1, K), \tag{5.69}$$

resulting

$$I_{\text{OUT}} = \frac{I_2}{4} - \frac{I_1}{2} + \frac{I_1^2}{4I_2}. \tag{5.70}$$

5.4.2.3 Implementation of the Multiplier/Divider Block

The implementation of the MD circuit that will be used for obtaining the function generated by the function synthesizer circuit is presented in Fig. 5.12, while its symbolical representation is shown in Fig. 5.13.

The expressions of I and I' currents are

$$I = \frac{I_C}{4} - \frac{I_A - I_B}{2} + \frac{(I_A - I_B)^2}{4I_C} \tag{5.71}$$

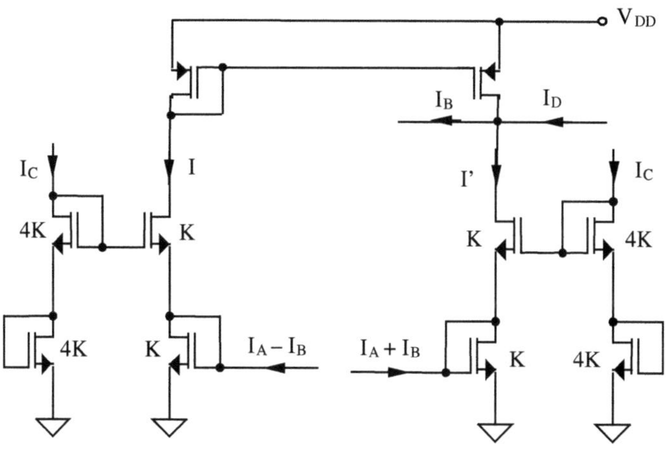

Fig. 5.12 Multiplier/divider circuit (*MD*) implementation

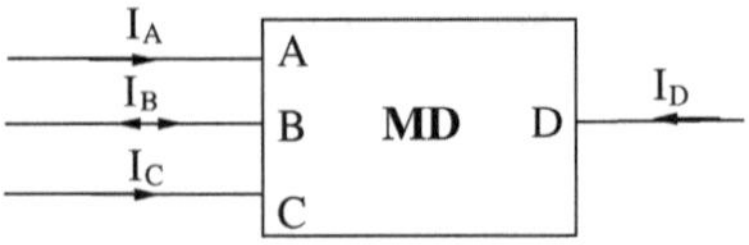

Fig. 5.13 Symbolical representation of the multiplier/divider circuit (*MD*)

and

$$I' = \frac{I_C}{4} - \frac{I_A + I_B}{2} + \frac{(I_A + I_B)^2}{4I_C}, \tag{5.72}$$

resulting

$$I_D = I' - I + I_B = \frac{I_A I_B}{I_C}. \tag{5.73}$$

5.4.2.4 Implementation of the Function Synthesizer

The function synthesizer implementation, based on the block diagram presented in Fig. 5.10 and on the MD circuit from Fig. 5.12, is presented in Fig. 5.14.

5.5 Fourth Function Synthesizer Circuit

5.5.1 Approximation Function

The fourth approximation function implements a linear combination of two primitive functions, $g_8(x)$ and $g_1(x)$:

$$g(x) = g_8(x) + cg_1(x) + dx + e, \tag{5.74}$$

a, *b*, *c*, *d*, and *e* being parameters determined from the condition of fourth-order approximation of $f(x)$ function using $g(x)$ function:

$$a = -\frac{(p+q)^3}{(r+q)(2q+r+p)}, \tag{5.75}$$

$$b = -\frac{q+r}{q+p}, \tag{5.76}$$

$$c = \frac{(q+p)^2}{2q+r+p} - p \tag{5.77}$$

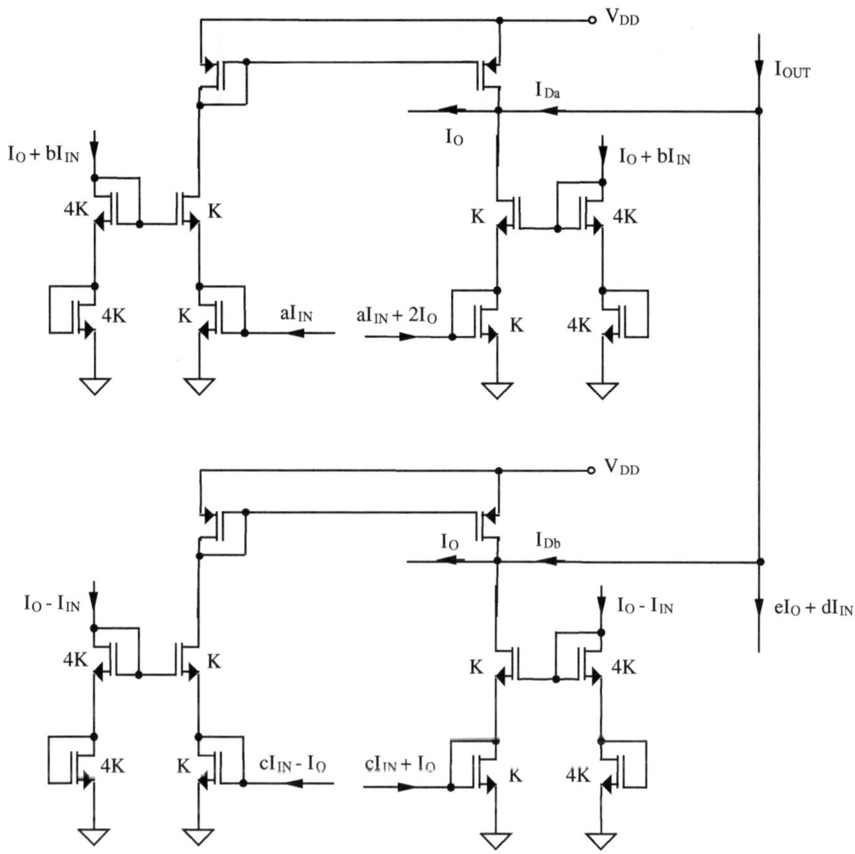

Fig. 5.14 Third function synthesizer circuit implementation

and

$$d = n + p + \frac{(q+p)^2}{q+r}, \tag{5.78}$$

$$e = m. \tag{5.79}$$

5.5.2 CMOS Implementation of the Function Synthesizer

5.5.2.1 Block Diagram of the Function Synthesizer

The fourth function synthesizer block diagram, using the fourth approximation function (5.74), is presented in Fig. 5.15.

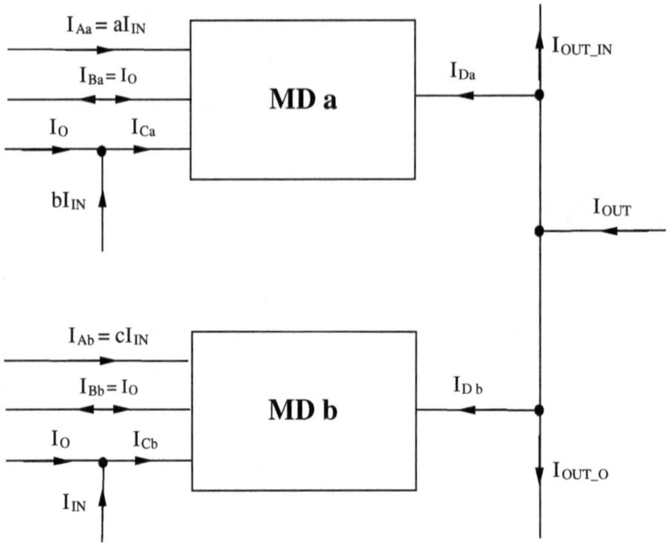

Fig. 5.15 Block diagram of the fourth function synthesizer

The "MD" structures have the implementation and description of their operation presented in the following paragraph. The I_{Da} and I_{Db} currents can be expressed as follows:

$$I_{Da} = \frac{I_{Ba} I_{Aa}}{I_{Ca}} \tag{5.80}$$

and

$$I_{Db} = \frac{I_{Bb} I_{Ab}}{I_{Cb}}. \tag{5.81}$$

The input currents of these blocks are:

$$I_{Aa} = aI_{\mathrm{IN}}, \tag{5.82}$$

$$I_{Ba} = I_O, \tag{5.83}$$

$$I_{Ca} = I_O + bI_{\mathrm{IN}}, \tag{5.84}$$

$$I_{Ab} = cI_{\mathrm{IN}}, \tag{5.85}$$

$$I_{Bb} = I_O \tag{5.86}$$

and

$$I_{Cb} = I_O + I_{\mathrm{IN}}. \tag{5.87}$$

The additional output currents can be expressed as follows:

$$I_{\mathrm{OUT\_IN}} = dI_{\mathrm{IN}} \tag{5.88}$$

and

$$I_{\mathrm{OUT\_}O} = eI_O. \tag{5.89}$$

The output current of the circuit having the block diagram presented in Fig. 5.15 will have the following expression:

$$I_{\mathrm{OUT}} = I_{Da} + I_{Db} + dI_{\mathrm{IN}} + eI_O. \tag{5.90}$$

Considering relation (5.74), I_{OUT} current represents the fourth-order approximation of $f(x)$ function:

$$I_{\mathrm{OUT}} = I_O\left(\frac{ax}{1+bx} + \frac{cx}{1+x} + dx + e\right) = I_O g(x) \cong I_O f(x). \tag{5.91}$$

5.5.2.2 Implementation of the Squaring Block

The core for implementing the function generator is represented by a current-mode SQ circuit, having the CMOS implementation presented in Fig. 5.16 [2].

The translinear loop that imposes the operation of the SQ circuit can be expressed as follows:

$$2V_{GS}(2I_2) = V_{GS}(I) + V_{GS}(I - I_1). \tag{5.92}$$

For a biasing in saturation of all MOS transistors from Fig. 5.23, it results

$$2\sqrt{2I_2} = \sqrt{I} + \sqrt{I - I_1}. \tag{5.93}$$

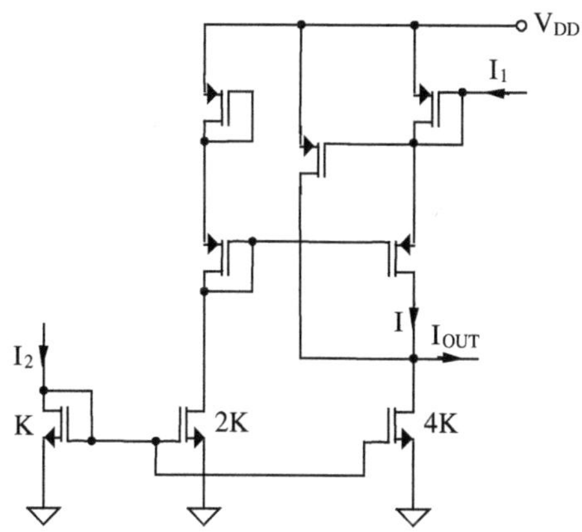

Fig. 5.16 Squaring circuit (*SQ*) implementation [2]

So, the I current will have the following expression:

$$I = 2I_2 + \frac{I_1}{2} + \frac{I_1^2}{32I_2}. \tag{5.94}$$

As a result of the circuit connections, the expression of I_{OUT} output current will be

$$I_{OUT} = I + (I - I_1) - 4I_2 = \frac{I_1^2}{16I_2}. \tag{5.95}$$

5.5.2.3 Implementation of the Multiplier/Divider Block

The MD circuit is presented in Fig. 5.17 [2], while its symbolical representation is shown in Fig. 5.18.

The expressions of I and I' currents are

$$I = \frac{(I_A + I_B)^2}{4I_C} \tag{5.96}$$

and

$$I' = \frac{(I_A - I_B)^2}{4I_C}, \tag{5.97}$$

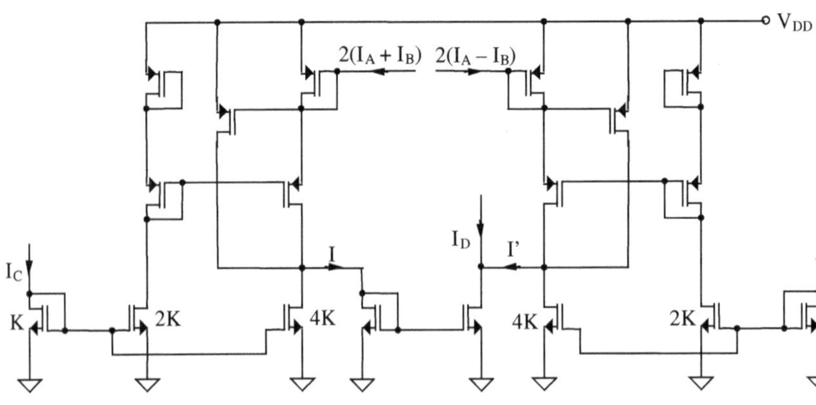

Fig. 5.17 Multiplier/divider circuit (*MD*) implementation [2]

Fig. 5.18 Symbolical representation of the multiplier/divider circuit (*MD*)

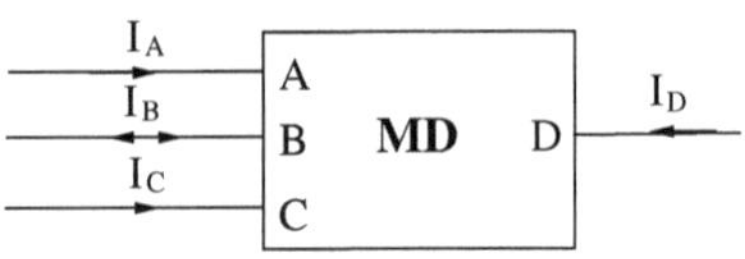

resulting

$$I_D = I - I' = \frac{I_A I_B}{I_C}. \tag{5.98}$$

5.5.2.4 Implementation of the Function Synthesizer

The implementation of the function synthesizer circuit, based on the block diagram presented in Fig. 5.15 and on the MD circuit from Fig. 5.17, is shown in Fig. 5.19 [2].

5.6 Fifth Function Synthesizer Circuit

5.6.1 Approximation Function

A linear combination of two primitive functions, $g_{10}(x)$ and $g_1(x)$, represents the basis for developing the fifth approximation function:

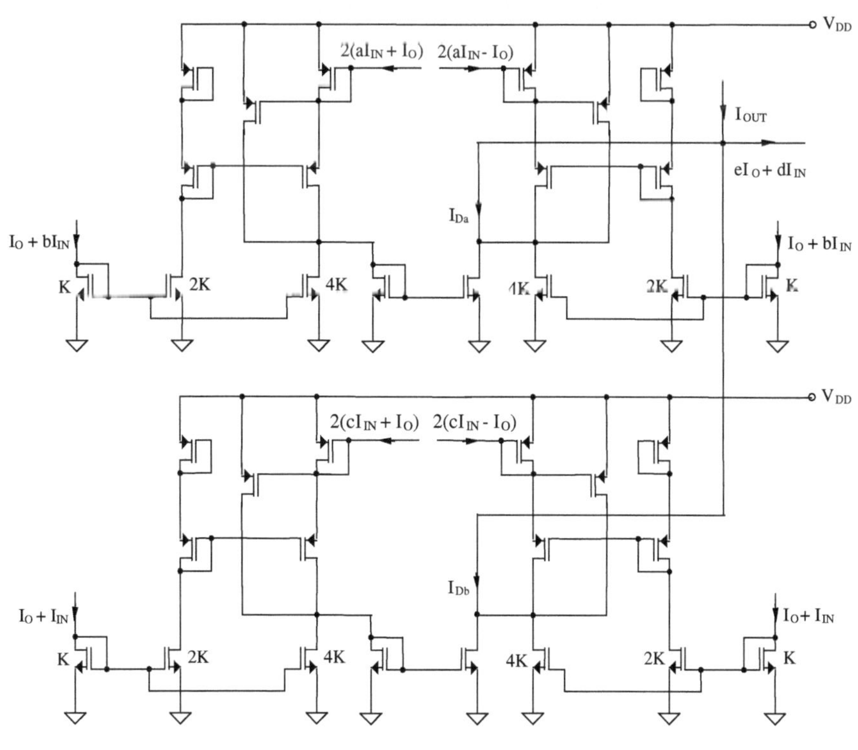

Fig. 5.19 Fourth function synthesizer circuit implementation [2]

$$g(x) = bg_{10}(x) + cg_{10}(x) + dx + e. \tag{5.99}$$

The a, b, c, d, and e parameters are determined from the condition of fourth-order approximation of $f(x)$ function using $g(x)$ function:

$$a = -\frac{(r+q)^2}{q^2 + rp + rq + pq}, \tag{5.100}$$

$$c = \frac{q^2 - rp}{r + p + 2q}, \tag{5.101}$$

$$b = \frac{p + c}{a^2}, \tag{5.102}$$

$$d = n - c + ab \tag{5.103}$$

and

$$e = m - b. \tag{5.104}$$

5.6.2 CMOS Implementation of the Function Synthesizer

5.6.2.1 Block Diagram of the Function Synthesizer

The block diagram of the fifth function synthesizer, based on the fifth approximation function (5.99), is presented in Fig. 5.20.

The "MD" and "SQ" circuits are current-mode MD and squaring structures, respectively, having implementation and description of their operation presented in the following paragraph. The I_D and I_G currents have the following expressions:

$$I_D = \frac{I_B I_A}{I_C} \tag{5.105}$$

and

$$I_G = \frac{I_E^2}{16 I_F} \tag{5.106}$$

where the input currents of these blocks are

$$I_A = cI_{\mathrm{IN}}, \tag{5.107}$$

$$I_B = I_O, \tag{5.108}$$

$$I_C = I_O + I_{\mathrm{IN}}, \tag{5.109}$$

$$I_E = I_O, \tag{5.110}$$

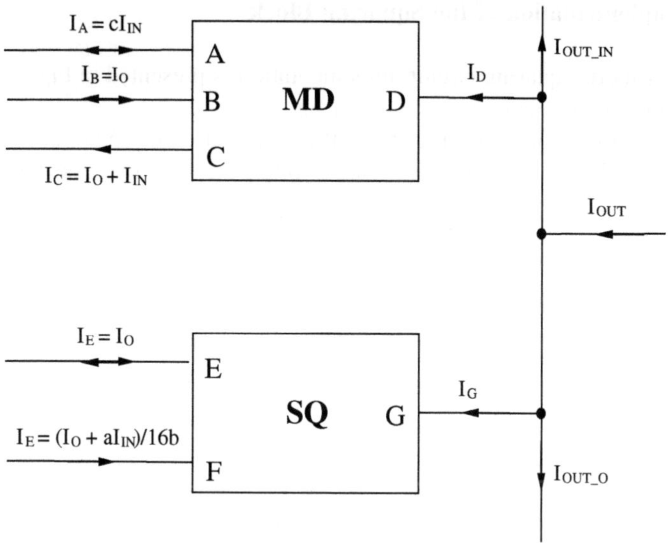

Fig. 5.20 Block diagram of the fifth function synthesizer

and

$$I_F = \frac{I_O + aI_{\text{IN}}}{16b}. \tag{5.111}$$

The expressions of additional output currents are:

$$I_{\text{OUT\_IN}} = dI_{\text{IN}} \tag{5.112}$$

and

$$I_{\text{OUT\_}O} = eI_O. \tag{5.113}$$

The output current of the circuit having the block diagram presented in Fig. 5.20 will have the following expression:

$$I_{\text{OUT}} = I_D + I_G + dI_{\text{IN}} + eI_O. \tag{5.114}$$

Using relation (5.99), it results that I_{OUT} current represents the fourth-order approximation of $f(x)$ function:

$$I_{\text{OUT}} = I_O\left(\frac{b}{1+ax} + \frac{cx}{1+x} + dx + e\right) = I_O g(x) \cong I_O f(x). \tag{5.115}$$

5.6.2.2 Implementation of the Squaring Block

The current-mode squaring circuit implementation is presented in Fig. 5.21, while its symbolical representation is shown in Fig. 5.22.

For MOS transistors from Fig. 5.21, the equation of the functional loop containing M1, M2, M3, and M4 gate-source voltages can be expressed as follows:

$$2V_{GS}(I_F) = V_{GS}(I) + V_{GS}(I + I_E), \tag{5.116}$$

resulting

$$2\sqrt{I_F} = \sqrt{I} + \sqrt{I + I_E} \tag{5.117}$$

So

$$I = I_F - \frac{I_E}{2} + \frac{I_E^2}{16I_F}. \tag{5.118}$$

The expression of the output current will be:

$$I_G = I - I_F + \frac{I_E}{2} = \frac{I_E^2}{16I_F}. \tag{5.119}$$

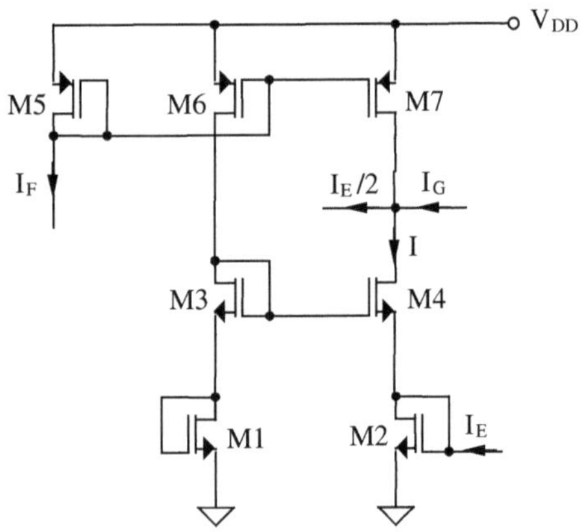

Fig. 5.21 Squaring circuit (*SQ*) implementation

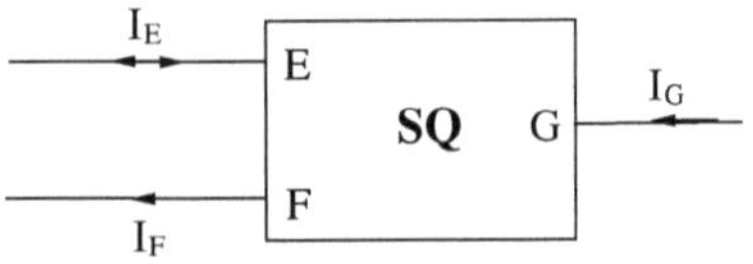

Fig. 5.22 Symbolical representation of the squaring circuit (*SQ*)

5.6.2.3 Implementation of the Multiplier/Divider Block

The implementation of the MD circuit that will be used for obtaining the function generated by the function synthesizer circuit is presented in Fig. 5.23, while its symbolical representation is shown in Fig. 5.24.

The I and I' currents can be expressed as follows:

$$I = \frac{(I_A - I_B)^2}{4I_C} \tag{5.120}$$

and

$$I' = \frac{(I_A + I_B)^2}{4I_C}, \tag{5.121}$$

resulting

$$I_D = I' - I = \frac{I_A I_B}{I_C}. \tag{5.122}$$

5.6.2.4 Implementation of the Function Synthesizer

The implementation of the function synthesizer circuit is presented in Fig. 5.25.

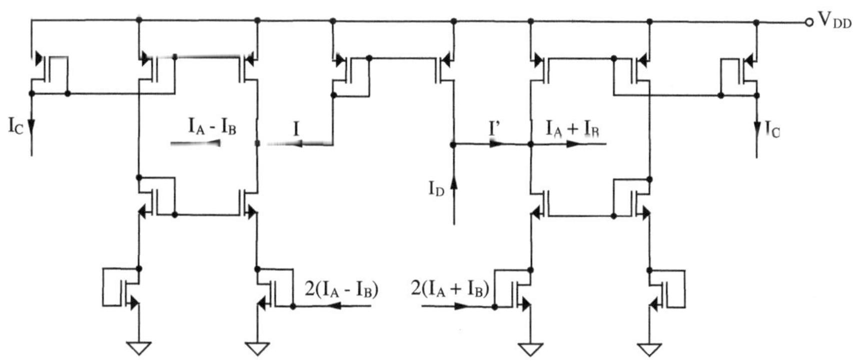

Fig. 5.23 Multiplier/divider circuit (*MD*) implementation

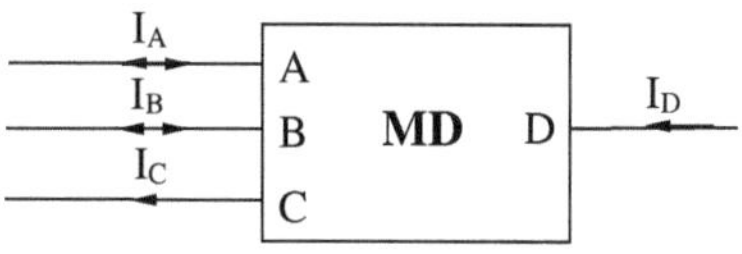

Fig. 5.24 Symbolical representation of the multiplier/divider circuit (*MD*)

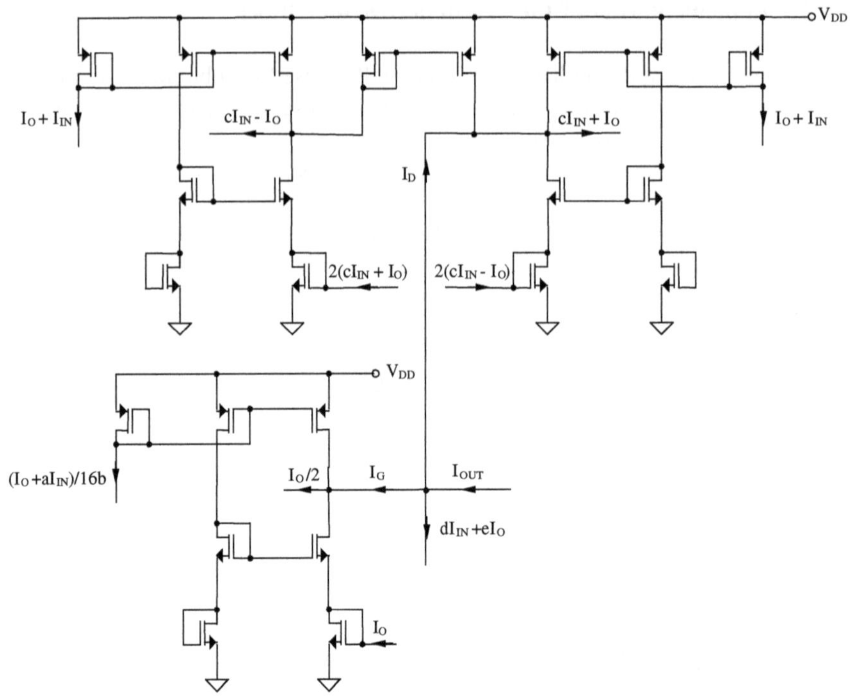

Fig. 5.25 Fifth function synthesizer circuit implementation

5.7 Sixth Function Synthesizer Circuit

5.7.1 Approximation Function

The sixth approximation function can be developed using a single primitive function, $g_{10}(x)$ and three additional squaring and linear terms:

$$g(x) = bg_{10}(x) + cx^2 + dx + e, \tag{5.123}$$

a, *b*, *c*, *d*, and *e* being parameters determined from the condition of fourth-order approximation of $f(x)$ function using $g(x)$ function:

$$a = -\frac{r}{q}, \tag{5.124}$$

$$b = \frac{q^4}{r^3}, \tag{5.125}$$

$$c = p - \frac{q^2}{r}, \tag{5.126}$$

$$d = n - \frac{q^3}{r^2} \tag{5.127}$$

and

$$e = m - \frac{q^4}{r^3}. \tag{5.128}$$

5.7.2 CMOS Implementation of the Function Synthesizer

5.7.2.1 Block Diagram of the Function Synthesizer

The block diagram of the seventh function synthesizer, based on the sixth approximation function (5.123), is presented in Fig. 5.26.

The "SQ" circuits are current-mode squaring circuits, having implementation and description of their operation presented in the following paragraph. The expressions of I_{Ga} and I_{Gb} currents are:

$$I_{Ga} = I_{Fb} - \frac{I_{Eb}}{2} + \frac{I_{Eb}^2}{16 I_{Fa}} \tag{5.129}$$

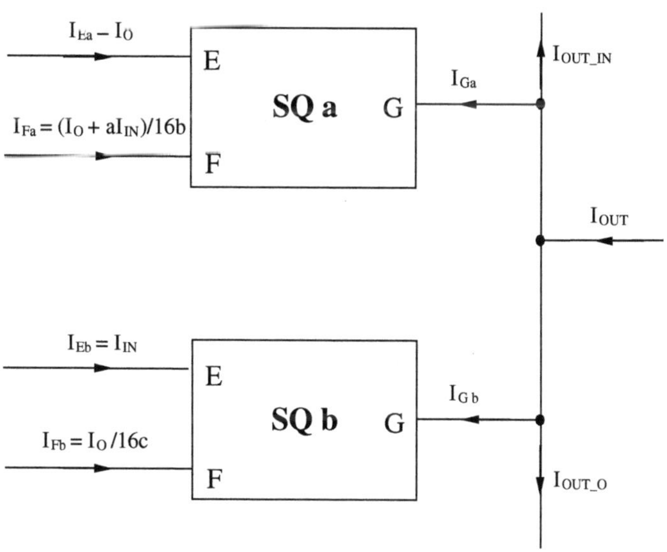

Fig. 5.26 Block diagram of the sixth function synthesizer

and

$$I_{Gb} = I_{Fb} - \frac{I_{Eb}}{2} + \frac{I_{Eb}^2}{16I_{Fb}}, \tag{5.130}$$

where the input currents of these blocks are

$$I_{Ea} = I_O, \tag{5.131}$$

$$I_{Fa} = \frac{I_O + aI_{\text{IN}}}{16b}, \tag{5.132}$$

$$I_{Eb} = I_{\text{IN}} \tag{5.133}$$

and

$$I_{Fb} = \frac{I_O}{16c}. \tag{5.134}$$

The additional output currents can be expressed as follows:

$$I_{\text{OUT\_IN}} = \left(\frac{1}{2} - \frac{a}{16b} + d\right) I_{\text{IN}} \tag{5.135}$$

and

$$I_{\text{OUT\_}O} = \left(\frac{1}{2} - \frac{a}{16b} - \frac{1}{16c} + e\right) I_O. \tag{5.136}$$

The output current of the circuit having the block diagram presented in Fig. 5.26 will have the following expression:

$$I_{\text{OUT}} = I_{Ga} + I_{Gb} + I_{\text{OUT\_IN}} + I_{\text{OUT\_}O}. \tag{5.137}$$

Taking into account relation (5.123), it results that I_{OUT} current represents the fourth-order approximation of $f(x)$ function:

$$I_{\text{OUT}} = I_O\left(\frac{b}{1 + ax} + cx^2 + dx + e\right) = I_O g(x) \cong I_O f(x). \tag{5.138}$$

5.7.2.2 Implementation of the Squaring Block

The core for implementing the function generator is represented by a current-mode squaring circuit, having the CMOS implementation presented in Fig. 5.27 [1], while its symbolical representation is shown in Fig. 5.28.

The equation of the translinear loop from Fig. 5.27 can be expressed as follows:

$$2V_{GS}(I_F) = V_{GS}(I_G) + V_{GS}(I_G + I_E), \tag{5.139}$$

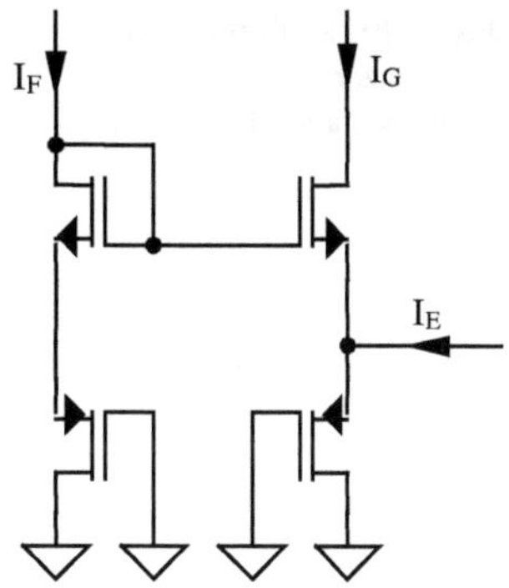

Fig. 5.27 Squaring circuit (*SQ*) implementation [1]

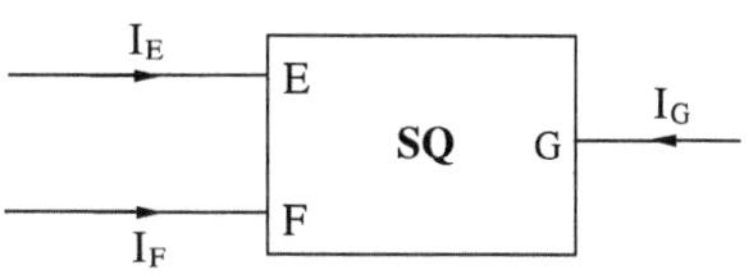

Fig. 5.28 Symbolical representation of the squaring circuit (*SQ*)

resulting

$$2\sqrt{I_F} = \sqrt{I_G} + \sqrt{I_G + I_E}. \tag{5.140}$$

So

$$I_G = I_F - \frac{I_E}{2} + \frac{I_E^2}{16I_F}. \tag{5.141}$$

5.7.2.3 Implementation of the Function Synthesizer

The implementation of the function synthesizer circuit, based on the block diagram presented in Fig. 5.28 and on the implementation of the SQ circuit from Fig. 5.27, is presented in Fig. 5.29 [1].

5.8 Seventh Function Synthesizer Circuit

5.8.1 Approximation Function

The seventh approximation function can be developed using a single primitive function, $g_8(x)$, and three additional linear and squaring terms:

$$g(x) = g_8(x) + cx^2 + dx + e. \tag{5.142}$$

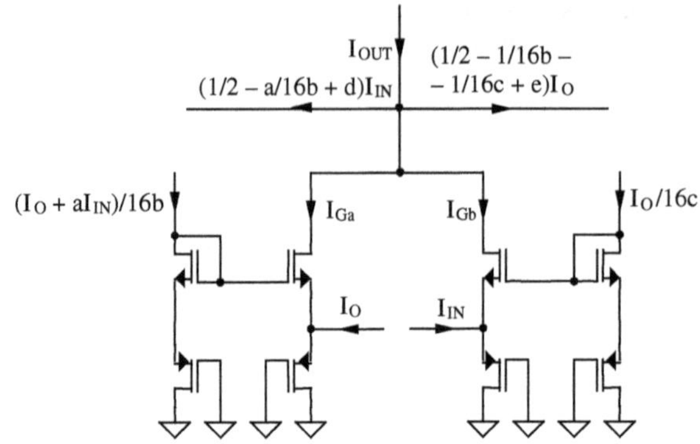

Fig. 5.29 Sixth function synthesizer circuit implementation [1]

The *a*, *b*, *c*, *d*, and *e* parameters are determined from the condition of fourth-order approximation of $f(x)$ function using $g(x)$ function:

$$a = \frac{q^3}{r^2}, \tag{5.143}$$

$$b = -\frac{r}{q}, \tag{5.144}$$

$$c = p - \frac{q^2}{r}, \tag{5.145}$$

$$d = n - \frac{q^3}{r^2} \tag{5.146}$$

and

$$e = m. \tag{5.147}$$

5.8.2 CMOS Implementation of the Function Synthesizer

5.8.2.1 Block Diagram of the Function Synthesizer

The block diagram of the seventh function synthesizer, based on the seventh approximation function (5.142), is presented in Fig. 5.30.

The expressions of I_D and I_G currents are

$$I_D = \frac{I_B I_A}{I_C} \tag{5.148}$$

and

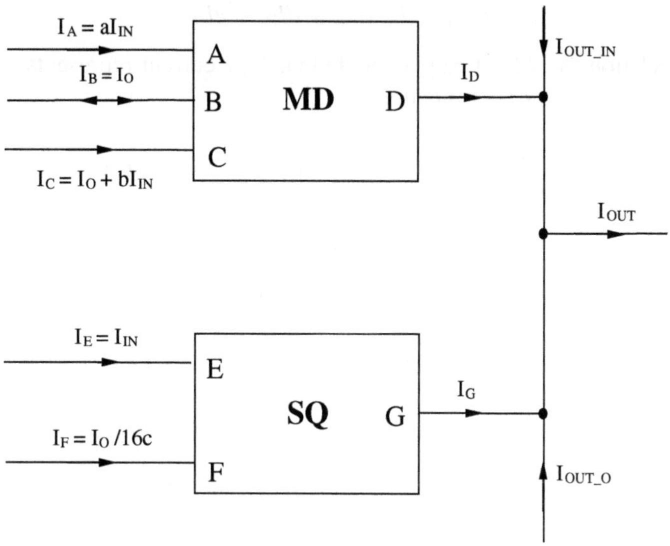

Fig. 5.30 Block diagram of the seventh function synthesizer

$$I_G = \frac{I_E^2}{16 I_F}. \tag{5.149}$$

where the input currents of these blocks are:

$$I_A = a I_{\text{IN}}, \tag{5.150}$$

$$I_B = I_O, \tag{5.151}$$

$$I_C = I_O + b I_{\text{IN}}, \tag{5.152}$$

$$I_E = I_{\text{IN}} \tag{5.153}$$

and

$$I_F = \frac{I_O}{16c}. \tag{5.154}$$

The additional output currents can be expressed as follows:

$$I_{\text{OUT\_IN}} = d I_{\text{IN}} \tag{5.155}$$

and

$$I_{\text{OUT\_}O} = e I_O. \tag{5.156}$$

The output current of the circuit having the block diagram presented in Fig. 5.30 will have the following expression:

$$I_{\mathrm{OUT}} = I_D + I_G + dI_{\mathrm{IN}} + eI_O. \tag{5.157}$$

Using relation (5.142), it can conclude that I_{OUT} current represents the fourth-order approximation of $f(x)$ function:

$$I_{\mathrm{OUT}} = I_O\left(\frac{ax}{1+bx} + cx^2 + dx + e\right) = I_O g(x) \cong I_O f(x). \tag{5.158}$$

5.8.2.2 Implementation of the Squaring Block (SQ)

The current-mode squaring circuit implementation is presented in Fig. 5.31, and its symbolical representation is shown in Fig. 5.32.

The equation of the translinear loop can be expressed as follows:

$$2V_{GS}(I_F) = V_{GS}(I_{D1}) + V_{GS}(I_{DL} + I_E), \tag{5.159}$$

resulting

$$2\sqrt{I_F} = \sqrt{I_{D1}} + \sqrt{I_{D1} + I_E}. \tag{5.160}$$

So

$$I_{D1} = I_F - \frac{I_E}{2} + \frac{I_E^2}{16I_F}. \tag{5.161}$$

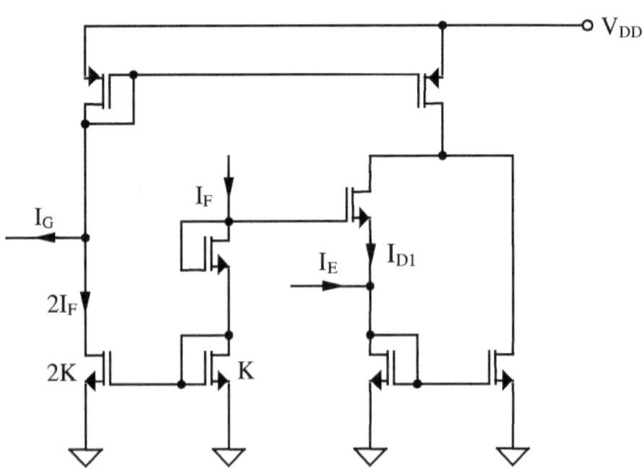

Fig. 5.31 Squaring circuit (*SQ*) implementation

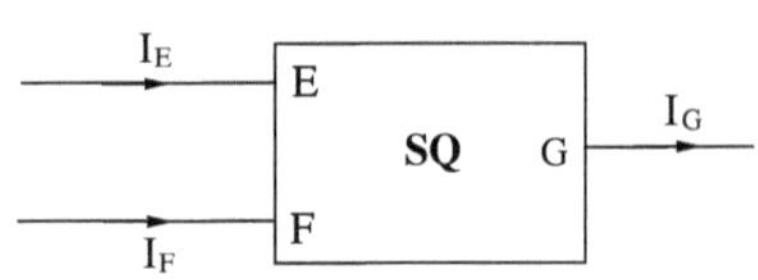

Fig. 5.32 Symbolical representation of the squaring circuit (*SQ*)

The expression of the output current of the SQ circuit will be

$$I_{OUT} = 2I_{D1} + I_E - 2I_F = \frac{I_E^2}{16I_F}. \tag{5.162}$$

5.8.2.3 Implementation of the Multiplier/Divider Block

The implementation of the MD circuit that will be used for obtaining the function generated by the function synthesizer circuit is presented in Fig. 5.33, while its symbolical representation is shown in Fig. 5.34.

The expressions of I and I' currents are

$$I = \frac{(I_A - I_B)^2}{4I_C} \tag{5.163}$$

and

$$I' = \frac{(I_A + I_B)^2}{4I_C}, \tag{5.164}$$

resulting

$$I_D = I' - I = \frac{I_A I_B}{I_C}. \tag{5.165}$$

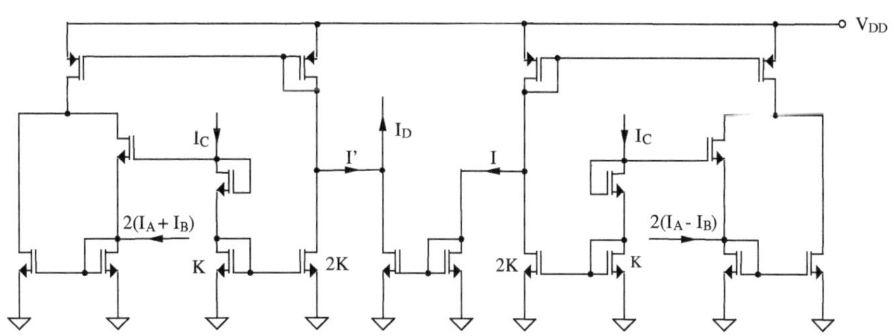

Fig. 5.33 Multiplier/divider circuit (*MD*) implementation

Fig. 5.34 Symbolical representation of the multiplier/divider circuit (*MD*)

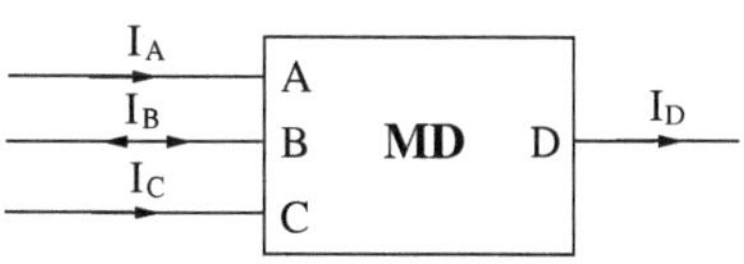

5.8.2.4 Implementation of the Function Synthesizer

The implementation of the function synthesizer circuit is presented in Fig. 5.35.

5.9 Eighth Function Synthesizer Circuit

5.9.1 Approximation Function

The eighth approximation function represents a linear combination of two primitive functions, $g_7(x)$ and $g_4(x)$:

$$g(x) = g_7(x) + cg_4(x) + dx + e, \tag{5.166}$$

a, b, c, d, and *e* being parameters determined from the condition of fourth-order approximation of $f(x)$ function using $g(x)$ function:

$$a = \frac{(p+q)^3}{(r+q)(r+p+2q)} - \frac{r+q}{p+q}, \tag{5.167}$$

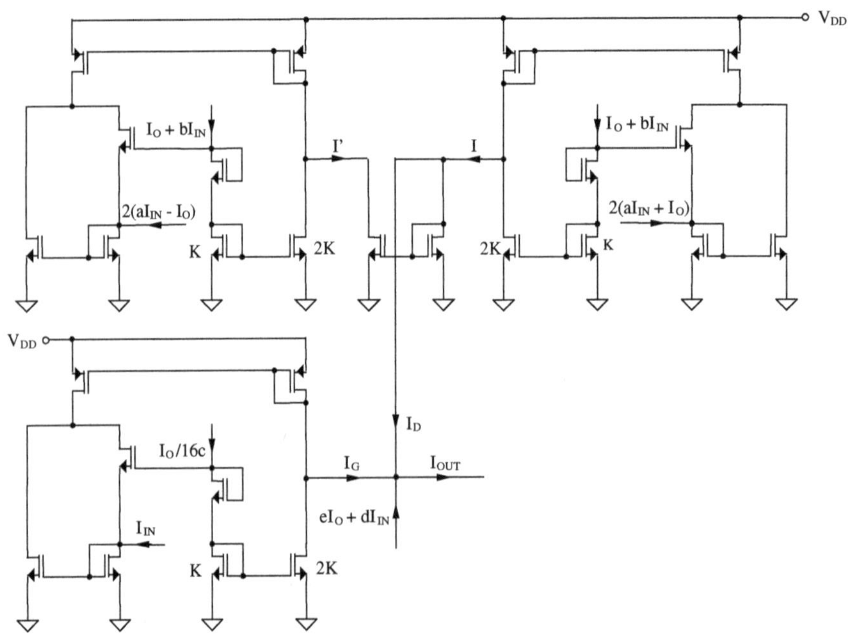

Fig. 5.35 Seventh function synthesizer circuit implementation

$$b = -\frac{r+q}{p+q}, \tag{5.168}$$

$$c = p - \frac{(p+q)^2}{r+p+2q} \tag{5.169}$$

and

$$d = n + p - \frac{(p+q)^2}{r+q}, \tag{5.170}$$

$$e = m - 1 - p + \frac{(p+q)^2}{r+p+2q}. \tag{5.171}$$

5.9.2 CMOS Implementation of the Function Synthesizer

5.9.2.1 Block Diagram of the Function Synthesizer

The block diagram of the eighth function synthesizer, using the eighth approximation function (5.166), is presented in Fig. 5.36.

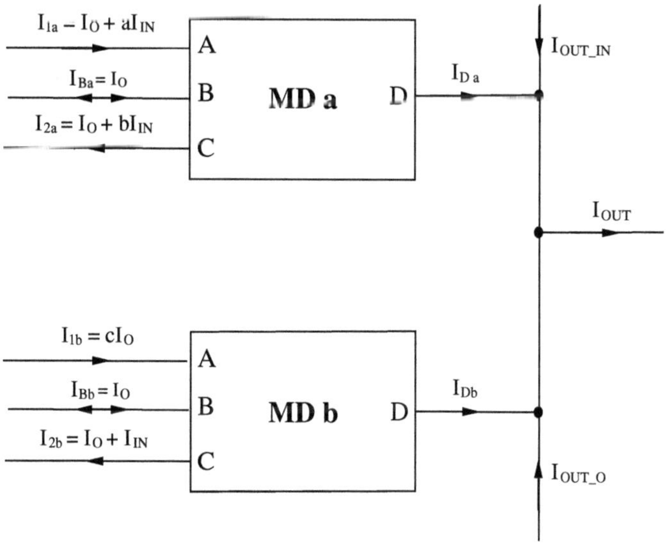

Fig. 5.36 Block diagram of the eighth function synthesizer

The I_{Da} and I_{Db} currents can be expressed as follows:

$$I_{Da} = I_{Ba} \frac{I_{Aa}}{I_{Ca}} \tag{5.172}$$

and

$$I_{Db} = I_{Bb} \frac{I_{Ab}}{I_{Cb}}. \tag{5.173}$$

The input currents of these blocks are

$$I_{Aa} = I_O + aI_{\mathrm{IN}}, \tag{5.174}$$

$$I_{Ba} = I_O, \tag{5.175}$$

$$I_{Ca} = I_O + bI_{\mathrm{IN}}, \tag{5.176}$$

$$I_{Ab} = I_{\mathrm{IN}}, \tag{5.177}$$

$$I_{Bb} = I_O \tag{5.178}$$

and

$$I_{Cb} = I_O + I_{\mathrm{IN}}. \tag{5.179}$$

The additional output currents have the following expressions:

$$I_{\mathrm{OUT-IN}} = dI_{\mathrm{IN}} \tag{5.180}$$

and

$$I_{\mathrm{OUT}-O} = eI_O. \tag{5.181}$$

The output current of the circuit having the block diagram presented in Fig. 5.36 will be expressed as follows:

$$I_{\mathrm{OUT}} = I_{Da} + I_{Db} + I_{\mathrm{OUT-IN}} + I_{\mathrm{OUT}-O}. \tag{5.182}$$

Considering relation (5.166), it results that I_{OUT} current represents the fourth-order approximation of $f(x)$ function:

$$I_{\mathrm{OUT}} = I_O \left(\frac{1+ax}{1+bx} + \frac{c}{1+x} + dx + e \right) = I_O g(x) \cong I_O f(x). \tag{5.183}$$

5.9.2.2 Implementation of the Multiplier/Divider Block

The MD circuit implementation is presented in Fig. 5.37, while its symbolical representation is shown in Fig. 5.38.

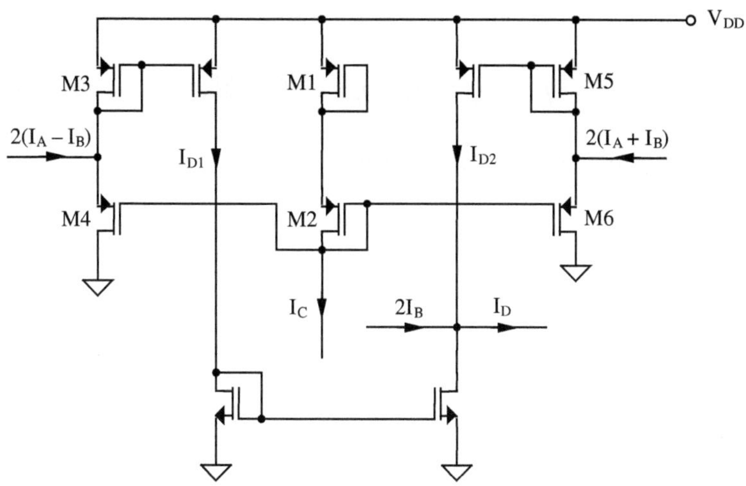

Fig. 5.37 Multiplier/divider circuit (*MD*) implementation

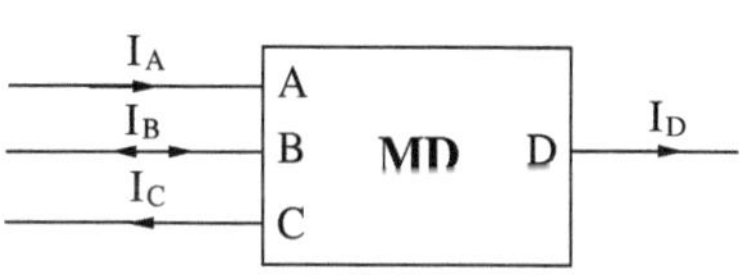

Fig. 5.38 Symbolical representation of the multiplier/divider circuit (*MD*)

The translinear loops containing M1–M4 and, respectively, M1, M2, M5, and M6 transistors implement the following relations:

$$V_{SG1} + V_{SG2} = V_{SG3} + V_{SG4} \tag{5.184}$$

and

$$V_{SG1} + V_{SG2} = V_{SG5} + V_{SG6}. \tag{5.185}$$

Biasing in saturation all identical MOS transistors from Fig. 5.37, it results

$$2\sqrt{I_C} = \sqrt{I_{D1}} + \sqrt{I_{D1} + 2(I_A - I_B)} \tag{5.186}$$

and

$$2\sqrt{I_C} = \sqrt{I_{D2}} + \sqrt{I_{D2} + 2(I_A + I_B)}, \tag{5.187}$$

resulting the following expressions of I_{D1} and I_{D2} currents:

$$I_{D1} = I_C - (I_A - I_B) + \frac{(I_A - I_B)^2}{4I_C} \tag{5.188}$$

and

$$I_{D2} = I_C - (I_A + I_B) + \frac{(I_A + I_B)^2}{4I_C}. \tag{5.189}$$

The output current of the "MD" circuits will be:

$$I_D = I_{D2} - I_{D1} + 2I_B = I_B \frac{I_A}{I_C}. \tag{5.190}$$

5.9.2.3 Implementation of the Function Synthesizer

The implementation of the function synthesizer circuit is presented in Fig. 5.39.

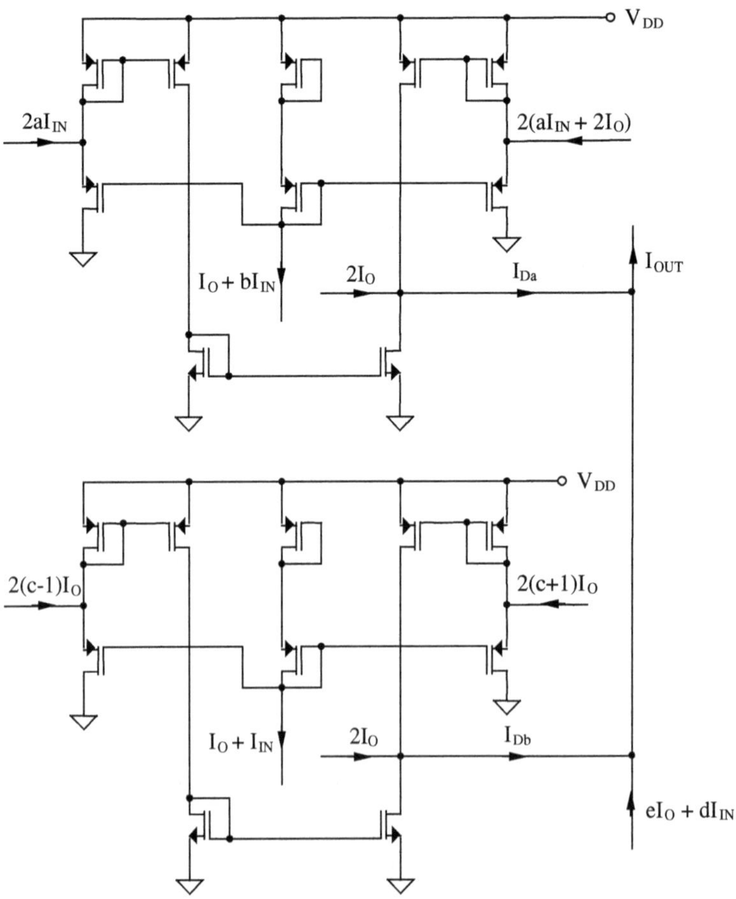

Fig. 5.39 Eighth function synthesizer circuit implementation

References

1. Cosmin Popa, "High-Accuracy Function Synthesizer Circuit with Applications in Signal Processing", EURASIP Journal on Advanced in Signal Processing, 2012
2. Cosmin Popa, "Low-Voltage Improved Accuracy Gaussian Function Generator with Fourth-Order Approximation", Microelectronics Journal, 2012, Volume 43 Issue 8, August, 2012, pp. 515-520

Index

DOI: 10.1007/978-3-319-01035-9,